HANDBOOK OF
Electronic Systems Design

Handbook of
Electronic Systems Design

FRANK WELLER

RESTON PUBLISHING COMPANY, INC.

A Prentice-Hall Company
Reston, Virginia

Library of Congress Cataloging in Publication Data
Weller, Frank.
  Handbook of electronic systems design.
  Includes index.
  1. Electronic apparatus and appliances—Design and
construction. 2. Systems engineering. I. Title.
TK7870.W36   621.381′042   77-10998
ISBN 0-87909-322-6

PRINTED IN THE UNITED STATES OF AMERICA

Contents

Preface ix

1 Introduction to System Design Principles 1

1–1 GENERAL CONSIDERATIONS 1
1–2 SYSTEM NOISE FACTORS 4
1–3 SYSTEM SIGNAL DISTORTION 7
1–4 CHARACTERISTIC IMPEDANCE 13
1–5 SYSTEM PHASE CHARACTERISTICS 17
1–6 SYSTEM BANDWIDTH-DISTORTION PRINCIPLES 21
1–7 GROUND CURRENTS IN TRANSMISSION SYSTEMS 23
1–8 TOLERANCES IN SYSTEM DESIGN 23
1–9 DESIGNING FOR SYSTEM RELIABILITY 26
1–10 SYSTEM DESIGN FOR MAXIMUM INFORMATION TRANSFER 28
1–11 TELEPHONE SYSTEM BASICS 30

2 Basic Radio Network Design 49

2–1 GENERAL CONSIDERATIONS 49
2–2 NETWORK CHANNELS AND STATION POWER 54
2–3 FM SYSTEM DESIGN CONSIDERATIONS 57
2–4 RADIO BROADCASTING SYSTEM 61
2–5 CABLE CARRIER SYSTEMS 66

3 Elements of Television System Design 70

3–1 GENERAL CONSIDERATIONS 70
3–2 SYSTEM DISTORTION PARAMETERS 73
3–3 NONLINEAR DISTORTION CHARACTERISTICS 86

4 Radar System Design Fundamentals 89

4–1 GENERAL CONSIDERATIONS 89
4–2 SYSTEM ORGANIZATION 92
4–3 TIMER AND TRANSMITTER 94
4–4 ANTENNA 96
4–5 RECEIVER FUNDAMENTALS 101

5 **Basic Telemetry Systems** 111

5–1 GENERAL CONSIDERATIONS 111
5–2 SYSTEM OVERVIEW 112
5–3 SIGNAL PROCESSING 116
5–4 PULSE CHARACTERISTICS 119
5–5 SENSORS AND SIGNAL CONDITIONERS 121
5–6 ANALOG-DIGITAL CONVERSION 124
5–7 TYPICAL PPM TELEMETRY SYSTEM 124
5–8 PDM TELEMETRY SYSTEM 131

6 **Elements of Microprocessor System Design** 135

6–1 GENERAL CONSIDERATIONS 135
6–2 FUNDAMENTAL MICROPROCESSOR CHARACTERISTICS 135
6–3 EXAMPLE OF EIGHT-BIT MICROPROCESSOR SYSTEM 147
6–4 INTEL 4040 MICROPROCESSOR SYSTEMS
6–5 INTEL 8080 MICROPROCESSOR SYSTEMS 168

7 **High-fidelity Systems** 178

7–1 GENERAL CONSIDERATIONS 178
7–2 BASIC AUDIO AMPLIFIER DESIGN 181
7–3 BASIC AMPLIFIER CHARACTERISTICS 184
7–4 COMPONENT AND DEVICE TOLERANCES 185
7–5 INPUT RESISTANCE VERSUS LOAD RESISTANCE 187
7–6 OUTPUT RESISTANCE VERSUS GENERATOR RESISTANCE 187
7–7 VOLTAGE AND CURRENT AMPLIFICATION 189
7–8 BASIC FET AMPLIFIER CHARACTERISTICS 192
7–9 PRINCIPLES OF NEGATIVE FEEDBACK 192
7–10 ACTIVE TONE CONTROL 200
7–11 POWER AMPLIFIER PARAMETERS 203
7–12 COMPLEMENTARY SYMMETRY AMPLIFIER OPERATION 204
7–13 SPEAKER SYSTEM CONSIDERATIONS 208
7–14 TONE-BURST TEST OF SPEAKER RESPONSE 214

8 **Community Antenna Television Systems** 217

8–1 GENERAL CONSIDERATIONS 217
8–2 FUNDAMENTALS OF RF CIRCUIT DESIGN 219
8–3 PRINCIPLES OF RESONANT CIRCUIT DESIGN 220
8–4 TRANSISTOR AND COUPLING NETWORK IMPEDANCES 228
8–5 TRANSFORMER COUPLING WITH TUNED PRIMARY 228
8–6 INTERSTAGE COUPLING WITH DOUBLE-TUNED NETWORKS 233
8–7 NEUTRALIZATION AND UNILATERALIZATION 240
8–8 LINE CHARACTERISTICS 245

9   Production Control Systems   253

9–1   GENERAL CONSIDERATIONS   253
9–2   STABILITY CRITERIA   256
9–3   WARD–LEONARD SERVO DRIVE   258
9–4   BASIC SERVO ARRANGEMENT   261
9–5   ELECTRONIC CONTROL OF BASIC SERVO SYSTEM   267
9–6   OTHER SERVOMECHANISM ARRANGEMENTS   267

Appendix I   269
Appendix II   272
Appendix III   273
Appendix IV   278
Index   281

Preface

With the rapid advance of electronics technology, an increasing need has arisen for a state-of-the-art electronic system design text. This book has been prepared to fill the gap between academic circuit theory and the practice of system design-engineering techniques. It will be found equally valuable in structured classroom instruction, in the design laboratory or shop, and in the home for self-instruction. Prerequisites include courses in basic electricity, electronics, and digital logic theory. It will also be helpful if the student has completed a course in elementary electronic circuit design. Although no knowledge of higher mathematics is required, the student should have completed courses in algebra, geometry, and trigonometry.

An introduction to system design principles is presented in the first chapter, with explanations and illustrations of basic system concepts. This discussion concludes with an overview of telephone system basics. The second chapter provides a basic treatment of radio network design. Problems of wave propagation and system-design approaches to communication reliability are explained. System organization is also covered, with its interrelations to telephone systems. In the third chapter, the elements of television system design are discussed and illustrated. Distinctions between radio system and video system engineering are pointed out. Basic test and maintenance requirements for video systems are included.

An introductory treatment of radar system design fundamentals is presented in the fourth chapter. Functions of subsections are explained and their relations to system operation are developed. In the fifth chapter, basic telemetry system design is discussed and illustrated. Principles of signal processing and conversion from analog to digital form, and vice versa, are explained. Basic circuitry for a complete system is included. The sixth chapter covers the elements of micro-processor system design. Because of their wide popularity, the types 4004, 8008, 4040, and 8080 microprocessors are detailed with respect to system operations.

High-fidelity systems are treated in the seventh chapter. Component type stereophonic and quadriphonic systems are illustrated. Basic audio design principles are described and illustrated. Both pre-amplifiers and power amplifiers are covered. The basics of speaker system design are included. In the eighth chapter, the essential features of community antenna television systems are discussed and exemplified. Basic high-frequency circuit, line, and network system principles are described. The ninth chapter introduces the student to the design of production control systems. Elements of automation are explained and illustrated. Basic servo systems and stability criteria are discussed. A detailed practical example of the Ward-Leonard servo system is included.

The author is indebted to his associates who have made many constructive criticisms and helpful suggestions. This text embodies a combination of dedicated effort, thousands of man-hours of experience in the educational field, and the perspective that results from team activity. It is appropriate that this text be dedicated as a teaching tool to the instructors and students of our junior colleges and community colleges, technical institutes, and vocational schools.

FRANK WELLER

HANDBOOK OF
Electronic Systems Design

1

Introduction to
System Design Principles

A system is an assembly of component parts linked together by some form of regulated interaction into an organized whole. Thus, the system designer defines a *television broadcast system,* a *community antenna television (CATV) system,* a *microprocessor system,* a *high-fidelity system,* a *radar system,* and so on. *System engineering* denotes a method of technical approach whereby all elements in a control system are considered, even to the smallest value and the process itself. A *system element* is one or more basic elements, together with other components necessary to form all or a significant part of one of the general functional groups into which a measurement system can be classified. From the viewpoint of operation, an important concept is *system deviation,* which denotes the instantaneous difference between the value of a specified system variable and the ideal value of the variable. Chart 1-1 tabulates some basic electronic systems.

    System reliability is also of basic concern to the system designer. This term denotes the probability that a system will perform its specified task properly under stated environmental conditions. Reliability can depend upon various factors. For example, in a communication system, reliability is dependent upon *system noise.* This factor is defined as the system output when it is operating with zero input signal. In a production control system that requires one or more manual operations, reliability is dependent in part upon *human engineering* factors. From the viewpoint of system engineering, these factors include functional indicator and control location, provision of ample visual contrast, and minimization of unnecessary distractions.

There is no sharp dividing line between circuits and systems. For example, a private telephone line between a design-engineering office and a production-control office is regarded as a *communication circuit.* On the other hand, a private branch exchange (PBX) in a manufacturing organization is termed a *manual telephone system.* Between these two extremes there is a "gray area" that can be regarded either as a complex communication circuit arrangement or as a comparatively simple communication system. A radio receiver or a radio transmitter is not regarded as a system. On the other hand, when these units are a functional part of a broadcast studio and field installation, as exemplified in Fig. 1-1, the receiver and transmitter are regarded as *subsystems.* In other words, a subsystem is a major, essential, functional part of a system.

The simplest type of communication system is generally considered to comprise five elements. An element is defined as any electrical device or component, such as an integrated circuit, transistor, transmission line, inductor, capacitor, resistor, antenna, transducer, and so on. Thus, a basic communication system comprises:

1. A source of information, such as a keyboard.
2. A coder for conversion of the information into a form suitable for transport by a transmission system or subsystem.
3. A means of signal transport, such as a cable or an electromagnetic wave transmitter.
4. A decoder for reconversion of the transmitted signal (pulses, for example) into a form suitable for actuating a reproducer.
5. A means of information reproduction, such as a printout unit.

1-2   SYSTEM NOISE FACTORS

Noise in an electronic system is defined as any unwanted disturbance, such as undesired electromagnetic radiation in a transmission channel, or as any unwanted electrical disturbance or spurious signal that modifies the transmitting, indicating, or recording of desired data. This definition includes random electrical variations that are internally generated by electronic devices or components. *Noise analysis* denotes a determination of the frequency components that are contained in the noise waveform. The electronic system designer also evaluates the *noise factor,* a ratio that is also termed the "noise figure." Within a stipulated

Chart 1-1 Basic electronic systems.

COMMUNICATION	CONTROL	GUIDANCE	DETECTION AND RANGING	DATA TRANSFER
Radio broadcast	Production operations	Autopilot	Radar	Telemetry
Television broadcast	Telecommunications	Missile	Sonar	Digital systems
Telephone	Space exploration	Space probes	Loran	
Commercial radio	TV transmitter remote control			
Satellite				
Microwave relay				
Outer space				
Community antenna television				
Intercommunication				

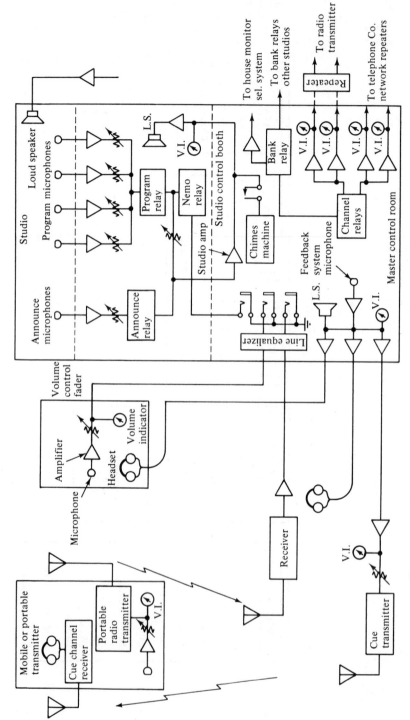

Figure 1-1 Example of a broadcast studio and field communication system.

4

transmission bandwidth, it is defined as the ratio of total noise at the output of the system to the noise at the input of the system.

Next, the *signal-to-noise ratio,* also called the "signal-noise ratio," is equal to the magnitude of the signal to that of the noise; it is commonly expressed in decibels. A signal-to-noise ratio must be specified with respect to a particular point of measurement in a system. In a television transmission system, the signal-to-noise ratio denotes the ratio in dB units of the maximum peak-to-peak voltage of the video signal (including the synchronizing pulses) to the voltage in rms units at a specified point of measurement. In a detector arrangement, a signal-to-noise ratio is equal to the amplitude of a signal after detection (demodulation) divided by the amplitude of the noise accompanying the signal. Note that a *noise level* denotes the amplitude of unwanted disturbances at a specified point of measurement in a system, customarily referred to as a *standardized level* and expressed as a power ratio in dB units.

Any transmission line or cable imposes progressive signal attenuation. If the signal becomes excessively attenuated, it will be masked by the noise level. Since it is impossible to recover the signal after it has become masked by noise voltages, amplifiers (boosters or repeaters) must be provided at suitable intervals along a transmission line. In general, it is desirable to keep the signal voltage at the highest possible level in a line, to reduce the possibility of noise interference. However, this is not always feasible. For example, if the signal level is raised substantially above 1500 μV in a community antenna television (CATV) cable, radiation leakage can become a problem and produce interference in TV receivers that are not connected to the cable. Again, in a commercial telephone system, crosstalk between channels places a practical limit on the maximum signal-voltage level that can be utilized.

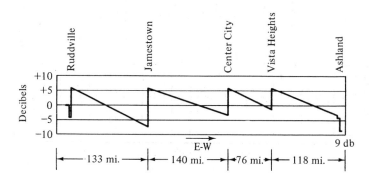

Figure 1-2   Signal level diagram for a typical transmission line.

6

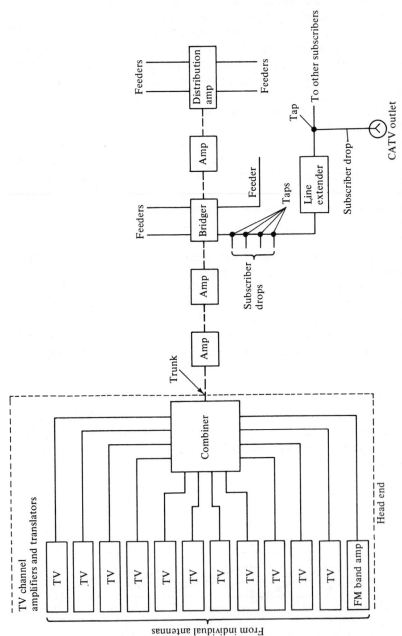

Figure 1-3 Arrangement of a typical CATV system.

Therefore, *a design compromise is employed wherein amplifiers are installed at various intervals along the transmission line.* These intervals are not critically spaced; however, they are chosen at sufficiently short intervals that there is no probability of the signal's becoming masked by noise, even under *worst-case* conditions. On the other hand, a tolerable minimum number of amplifiers is specified from the viewpoint of system costs. A signal level diagram for a typical telephone transmission line is shown in Fig. 1-2. Observe that the signal at Ruddville is 5 dB below reference level, and is elevated to 6 dB above reference level by a repeater. After 133 miles of transmission it arrives at Jamestown with a level that is now 7 dB below reference level. Thereupon, it is elevated to 6 dB above reference level by a repeater, after which it progresses to Center City. Note in passing that the signal-to-noise ratio deteriorates somewhat during each interval, and that *amplification cannot improve the input signal-to-noise ratio.* As an illustration of the basic problem confronted by the system designer, the loop resistance of a telephone line between New York and San Francisco is approximately 500,000 ohms (0.5 megohm).

Next, observe the arrangement of a typical CATV system, depicted in Fig. 1-3. From the combiner network, the various signal voltages are fed into a trunk cable. A trunk cable serves to conduct the signal voltages from their origin, such as atop a mountain, to the site of utilization, such as a residential area several miles away. Because the video and sound signals are progressively attenuated through a trunk line, amplifiers must be inserted at intervals. Otherwise, the signal-to-noise ratio would deteriorate to a point that reproduced pictures would be snowy. After a TV signal has become snowy, no amount of amplification can subsequently improve the signal-to-noise ratio. A typical coaxial cable imposes an attenuation of 1 dB/100 ft of cable at an operating frequency of 150 MHz. In turn, an amplification of approximately 50 dB/mile is required.

1-3 SYSTEM SIGNAL DISTORTION

The electronic system designer is also concerned with system signal distortion. This is a procedure termed *network analysis,* for determination of the electrical properties of a network. For example, its input and transfer impedances, responses, and other characteristics are derived from the configuration, parameters, and driving voltages of the network. Note that from a circuit viewpoint, a network is regarded as a combination of electrical elements. On the other hand, from a system viewpoint, a network is defined as an interconnected system of trans-

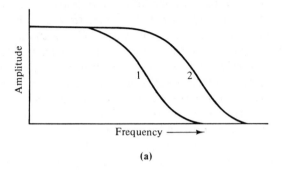

(a)

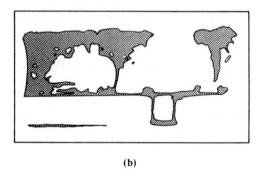

(b)

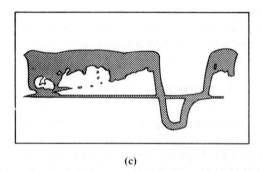

(c)

Figure 1-4 Sync-pulse distortion caused by limited bandwidth in a video-signal system. (a) Narrow bandwidth (1), and wide bandwidth (2); (b) pulse transmission through wide-band system; (c) pulse transmission through narrow-band system.

mission lines that provides multiple connections between sources of signals and remote loads. As noted above, a network generally includes amplifiers, and system responses may depend considerably upon amplifier characteristics. As an illustration, Fig. 1-4 shows how a synchronizing pulse becomes distorted by transmission through a narrow-band system.

Circuit designers are familiar with the concept of transconductance—the ratio of a change in output current to the corresponding change of input voltage. Similarly, system designers employ a related concept called the *transfer impedance* of a network. It is defined as the ratio of the potential difference between any two pairs of terminals of a network applied at one pair of terminals to the resultant current at the other pair of terminals (all terminals being terminated in any specified manner). Accordingly, *a nonlinear transfer impedance results in nonlinear distortion of a signal during transmission from the input terminals to the output terminals of the network.* Note in passing that the *transfer admittance* of a network is equal to the reciprocal of its transfer impedance.

Basic types of system signal distortion are termed:

1. Nonlinear distortion that develops because the transmission properties of a system are dependent upon the instantaneous amplitude of the transmitted signal. It is characterized by incremental amplitude and harmonic distortion, intermodulation, and flutter.
2. Frequency distortion that develops when all frequencies in a complex waveform are not attenuated equally, or are not amplified equally, in passage through a network.
3. Delay distortion, also called phase-delay distortion, denotes any departure from flatness in the phase delay characteristic of a network over the frequency range required for the transmission; it can also be regarded as the *effect* of such departure on a transmitted signal. Delay distortion can be conveniently evaluated as the amount of variation in the delay imposed upon various frequency components of the signal, expressed in microsecond units with respect to an average delay time.

It is instructive to consider the change in resultant waveshape that results from delay distortion in basic fundamental and single-harmonic waveforms. Typical complex waveforms resulting from a mixture of a fundamental frequency with a second-harmonic, or third-harmonic, or fourth-harmonic, or fifth-harmonic frequency, are illustrated in Fig. 1-5. Observe that the waveshape changes considerably as a result of

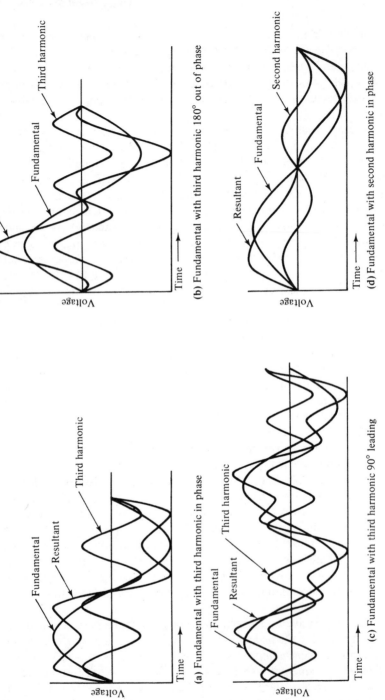

Figure 1-5 Basic fundamental and single harmonic waveforms.

(a) Fundamental with third harmonic in phase

(b) Fundamental with third harmonic 180° out of phase

(c) Fundamental with third harmonic 90° leading

(d) Fundamental with second harmonic in phase

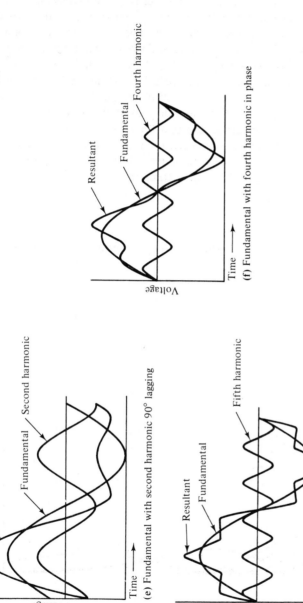

(e) Fundamental with second harmonic 90° lagging

(g) Fundamental with fifth harmonic in phase

(f) Fundamental with fourth harmonic in phase

Figure 1-5 *Continued* **Basic fundamental and single harmonic waveforms.**

a shift in phase of a harmonic with respect to its fundamental component. Phase relations are directly related to coincidence in time, or its lack (delay of one frequency component relative to another component). Note that odd harmonics always produce symmetrical positive and negative excursions. Observe in Fig. 1-5(d) that in-phase second harmonic does not produce a symmetrical resultant. In other words, the negative excursion is a mirror image of the positive excursion, and is therefore unsymmetrical. Note also that all of the waveforms shown in Fig. 1-5 are AC waveforms, and that their average value is zero in each case. When a harmonic is delayed (shifted in phase), the resultant waveform still has an average value of zero. That is, it is impossible to form a complex wave that has a DC component from any mixture of sine waves.

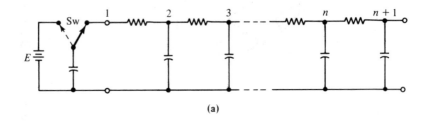

(a)

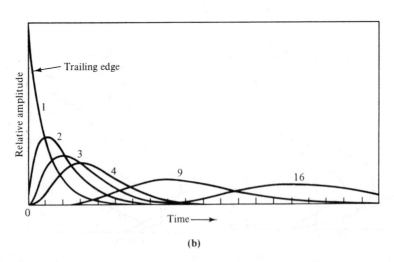

(b)

Figure 1-6   Responses produced by successive integrating circuit sections to an applied pulse. **(a)** RC integrating network; **(b)** pulse is progressively attenuated, stretched, and delayed.

A *network analyzer* is a group of circuit elements that can be interconnected to form network models. From corresponding measurements on the model, it is then possible to infer the electrical parameters at various points in the prototype system. For example, a transmission line can be roughly simulated in the first analysis by a series of RC sections, as depicted in Fig. 1-6. When a pulse voltage from the input charged capacitor is applied to the ladder network, the pulse is propagated toward the output end of the ladder. As each L section is traversed, the pulse becomes progressively attenuated, stretched, and delayed. In other words, the pulse width becomes progressively greater. An L section comprises a parallel element followed by a series element, or vice versa. The pulse amplitude decreases because of I^2R loss in successive series resistors. Because each successive capacitor must be charged over a finite interval from the preceding capacitor, the appearance of the pulse becomes progressively delayed at each successive line section.

1-4 CHARACTERISTIC IMPEDANCE

Characteristic impedance, also termed *surge impedance,* is defined as the input impedance of a line if it were of infinite length. In a *delay line,* the characteristic impedance is the value of terminating resistance that provides minimum reflection to the network input and output signals. The characteristic impedance is also presented as the ratio of voltage to current at every point along a transmission line on which there are no reflections (standing waves). In the case of a resistive ladder, there is no reactive component, and the system designer speaks of the *characteristic resistance* of the network. A development of the characteristic resistance concept is shown in Fig. 1-7. In this example, 10-ohm series resistors and 100-ohm shunt resistors are connected into a progressively longer ladder. If one L section is checked for input resistance, a current of 0.91 A is drawn in response to the application of 100 volts; the input resistance is 110 ohms. When six L sections are included, a current of 2.59 A is drawn with an input voltage of 100 V; the input resistance is 38.6 ohms.

Next, it will be shown that the value of the input resistance to the foregoing ladder is approaching 37 ohms, approximately, as a limit, as noted in Fig. 1-7(h). Refer to Fig. 1-8(a). Since infinity minus one is still equal to infinity, a single L section can be analyzed for the characteristic resistance value. Thus, the L section comprising R_1 and R_2 is terminated by R_0, and it is stipulated that input resistance to the

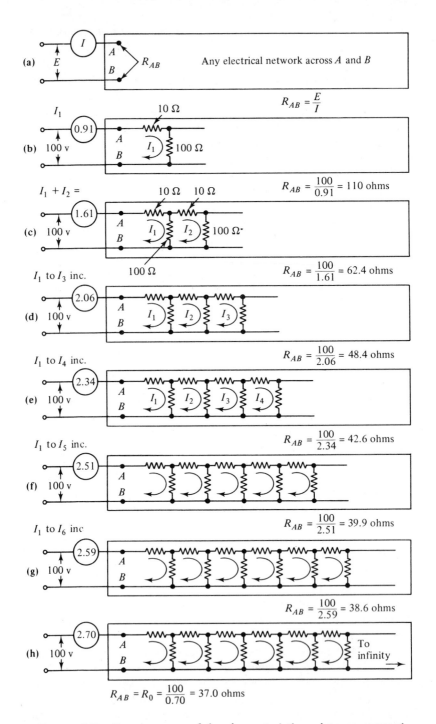

Figure 1-7 Development of the characteristic resistance concept.

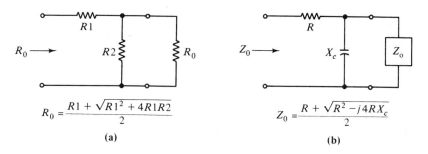

$$R_0 = \frac{R1 + \sqrt{R1^2 + 4R1R2}}{2}$$

(a)

$$Z_0 = \frac{R + \sqrt{R^2 - j4RX_c}}{2}$$

(b)

Figure 1-8 Characteristic resistance and impedance of ladder networks. (a) Resistive ladder arrangement; (b) resistance-capacitance ladder arrangement.

terminated L section is equal to R_0. In turn, solution of the resulting quadratic formula gives a general expression for R_0, and when $R_1 = 10$, and $R_2 = 100$, this characteristic-resistance value is 37 ohms, approximately. Similarly, the characteristic impedance of the RC ladder depicted in Fig. 1-6(a) can be analyzed as shown in Fig. 1-8(b). This characteristic impedance is not a constant, but is a function of frequency; it consists of a resistive component and a capacitive component.

Coaxial cable is extensively employed by system designers as a transmission line. Short lengths (stubs) of coaxial cable with a short-circuited termination, or an open-circuited termination, are also utilized as series-resonant, parallel-resonant, and reactive elements at high frequencies. A short length of cable can be regarded as lossless for nearly all practical purposes. In turn, a "stub" has a characteristic resistance expressed by the equation:

$$R_0 = \sqrt{\frac{L}{C}}$$

In the foregoing equation, L denotes the amount of distributed inductance per unit length of cable. Distributed inductance is a continuous parameter, in contrast to the concentrated or lumped inductance of a coil. Similarly, C denotes the amount of distributed capacitance per unit length of cable, in the foregoing equation. It is a continuous parameter, in contrast to the concentrated or lumped capacitance of a conventional capacitor. Typical values for R_0 fall in the range from 50 to 75 ohms. However, the system designer occasionally utilizes cables that have higher or lower values of characteristic impedance.

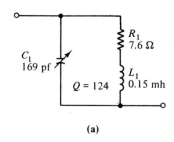

(a)

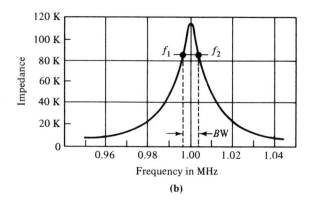

(b)

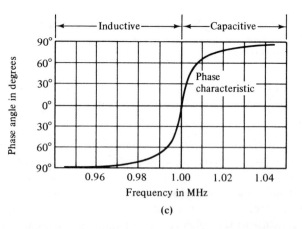

(c)

Figure 1-9 Frequency-response curve and phase characteristic for a parallel-resonant circuit. **(a)** Configuration; **(b)** frequency response; **(c)** phase characteristic.

1-5 SYSTEM PHASE CHARACTERISTICS

It follows from the previous discussion that signal (waveform) distortion can be caused by an improper phase characteristic of a system. It is helpful to recall that a tuned circuit has both a specified frequency response and an associated phase characteristic, as shown in Fig. 1-9. With reference to this parallel-resonant circuit, the fundamental parameters to be noted are:

1. The tuned circuit "looks like" a pure resistance at 1 MHz, because the phase angle is zero at the peak of the frequency response curve.
2. At frequencies below 1 MHz, the tuned circuit "looks like" a coil with a resistive component. (Lagging phase angle.)
3. At frequencies above 1 MHz, the tuned circuit "looks like" a capacitor with a resistive component. (Leading phase angle.)
4. At the 70.7 percent-of-maximum points on the frequency-response curve (half-power points), the tuned circuit "looks like" a resistance connected in series with an equal value of reactance. (Phase angle of 45 deg.)
5. The phase characteristic is not linear (not a straight line). This phase characteristic is particularly nonlinear in the regions of the half-power frequencies on the frequency-response curve.

Next, observe the idealized frequency and phase characteristics for a TV signal transmission system, shown in Fig. 1-10. The frequency response is uniform (flat) from zero frequency (DC) to the cutoff frequency of the system. The *phase characteristic is linear* (a straight line), and it *passes through the origin*. This is called a *uniform time-delay characteristic*. Note carefully that:

1. *The slope of the phase characteristic is equal to the time delay of a signal that passes through the system from input to output.* This slope is equal to a chosen phase-angle interval divided by the corresponding frequency interval.
2. *Time delay is proportional to the product of phase angle and frequency.* Thus, if a delay of 12 μs equals 35 deg at 0.01 MHz, it also equals 350 deg at 0.1 MHz.

To summarize briefly, a phase characteristic is a graph of phase shift versus frequency, for the condition of sinusoidal input and output in the system. In turn, phase-delay distortion is based on the difference between the phase delay at one frequency and the phase delay at a

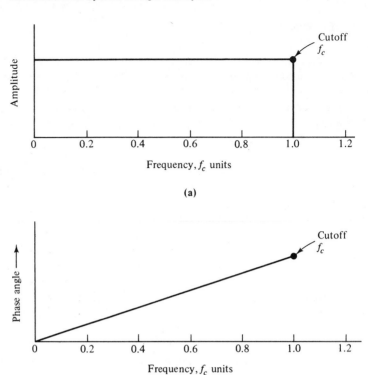

Figure 1-10   Idealized frequency and phase characteristics for a TV signal transmission system. **(a)** Frequency response; **(b)** phase characteristic.

reference frequency. A square wave can be regarded as containing an array or spectrum of sine-wave components, as depicted in Fig. 1-11. The fundamental frequency of the square wave is its reference frequency. It is sometimes called the first harmonic; the square waveform consists of this fundamental frequency, a third-harmonic frequency, a fifth-harmonic frequency, and so on. All frequency components are in phase; that is, they pass through the origin at the same time. It follows that if this square-wave frequency spectrum passes through a system that delays its frequency components disproportionately, the output waveform will be distorted, and will not be a true square wave. One basic aspect of phase distortion is a sloping top in a reproduced square wave, as depicted in Fig. 1-12.

An *equalizer* is a passive device that is designed to compensate for an undesired amplitude-frequency and/or phase-frequency charac-

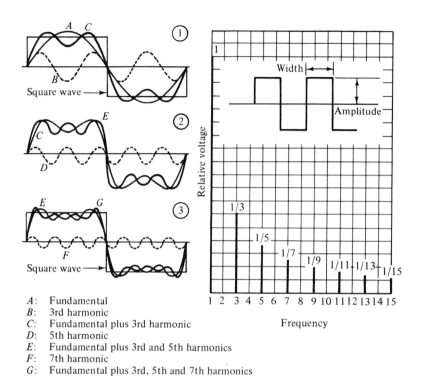

A: Fundamental
B: 3rd harmonic
C: Fundamental plus 3rd harmonic
D: 5th harmonic
E: Fundamental plus 3rd and 5th harmonics
F: 7th harmonic
G: Fundamental plus 3rd, 5th and 7th harmonics

(a)   (b)

Figure 1-11   Synthesis of a square wave from harmonically related sine waves.

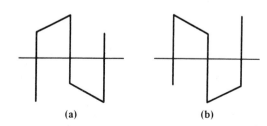

(a)   (b)

Figure 1-12   Examples of phase distortion in a reproduced square wave. **(a)** Low-frequency lag; **(b)** low-frequency lead.

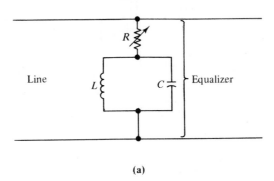

(a)

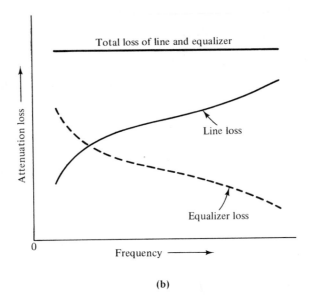

(b)

Figure 1-13 A simple bridged frequency equalizer. **(a)** Configuration; **(b)** frequency response.

teristic of a system. Thus, the designer employs both frequency equalizers and phase equalizers. An example of a frequency equalizer is shown in Fig. 1-13. This *LCR* configuration is designed to have an attenuation characteristic that is the opposite of the line attenuation characteristic. In turn, the effective frequency response of the line becomes uniform. On the other hand, the frequency equalizer does not correct the phase characteristic of the line, although it changes the original phase characteristic. In other words, a separate phase equalizer

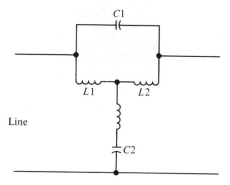

Figure 1-14   Example of a bridged-T type of phase equalizer.

must be utilized to compensate for any undesired phase-frequency characteristic.

An example of a phase-equalizer configuration is shown in Fig. 1-14. This is called a bridged-T type of equalizer, because capacitor C1 is bridged across L1 and L2. By way of comparison, the frequency-equalizer configuration depicted in Fig. 1-13 is not a bridged-T arrangement. However, it is called a bridged equalizer because it is bridged across the line. Note that when a phase equalizer is added to a system that contains a frequency equalizer, the net frequency response will no longer be uniform. In other words, a phase equalizer interacts with a frequency equalizer, and their component values must be modified accordingly in order to obtain a flat frequency response and a linear phase characteristic simultaneously. System designers often employ two or more frequency equalizers connected in cascade, and two or more phase equalizers connected in cascade.

1-6   SYSTEM BANDWIDTH-DISTORTION PRINCIPLES

A transmission system may have an essentially linear phase characteristic, although its bandwidth is limited. In such a case, the higher harmonics of a square wave will become attenuated in passage of the waveform through the system. In turn, the output waveform will exhibit frequency distortion with negligible phase distortion. Frequency distortion alone introduces corner-rounding in the reproduced square wave, but it does not develop any tilt along the top and bottom excursions of the waveform. Frequency distortion (attenuation of the higher harmonics) also slows down the rise time and the fall time in a square

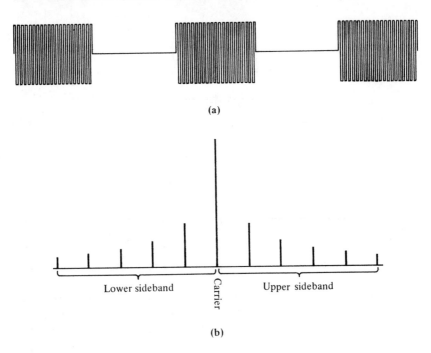

Figure 1-15 Carrier 100 percent amplitude-modulated by a square wave. **(a)** Modulated waveform; **(b)** partial frequency spectrum.

wave. To cite an extreme case, if all of the harmonics were removed from a square wave, and only the fundamental component proceeded to the output of the system, the rise time and the fall time of the output waveform would have their maximum values, and the "corners" of the waveform would be changed into a sinusoidal form.

Consider next the passage through a transmission system of a carrier frequency that is amplitude-modulated by a square wave. With reference to Fig. 1-15, the modulated waveform consists of a frequency spectrum that contains two sidebands. Note that each sideband has a structure that is the same as the square-wave frequency spectrum depicted in Fig. 1-11(b). Consequently, the transmission channel must have twice the bandwidth of a channel that is utilized to transport a corresponding unmodulated square-wave frequency spectrum. It is evident that since the same basic frequency spectrum is transported in either case, *that channel-bandwidth frequency limitation will have the same fundamental distorting action on a modulated waveform as on its unmodulated counterpart.*

1-7 GROUND CURRENTS IN TRANSMISSION SYSTEMS

The soil in most geographic areas can be used as a conductor in an electric circuit, and as a means of dissipating excessive charges caused by lightning or nearby power lines. Connections to the earth, called *grounds* or *ground connections,* are made to pipes of buried water systems, driven rods, buried metal plates, or buried wire. In most locations, the use of a circuit ground in a telephone communication system induces noise in the circuits and causes crosstalk with other channels that utilize circuit grounds in the vicinity. Therefore, it is very poor practice to employ a circuit ground. On the other hand, equipment grounds are used in all communication systems to drain off static disturbances. The purpose of a ground connection is to keep some point in a system as near as possible at ground potential. In many instances it is necessary that there be a considerable flow of current through a ground connection to prevent the potential of the connection from rising to a value above actual earth potential. The soil offers more or less resistance to current flow, and this resistance determines in large measure the effectiveness of a ground connection.

As a general rule, a potential difference exists between any two ground connections that are separated by a substantial distance. This potential fluctuates in most cases. When factory ground systems are involved, for example, an appreciable 60-Hz potential fluctuation is often established between two ground systems that are located some distance apart. In this example, the 60-Hz potential difference is attributable to different ground currents from the electrical systems of the two factories. With reference to Fig. 1-16, the outer conductor of the coaxial cable has much less resistance than that of the soil in which it is buried. Consequently, the 60-Hz potential difference between the ends of the cable causes a 60-Hz current flow through its outer conductor. Since this conductor resistance is not zero, a 60-Hz voltage drop is established between the ends of the cable. This voltage is termed *hum voltage,* and it mixes with the signal voltage that is transported by the cable.

1-8 TOLERANCES IN SYSTEM DESIGN

System designers are often concerned with tolerances on signal characteristics, whereas circuit designers usually start with precise input signal voltages and waveforms. For example, a television network must transport and process signals from many locations. In some cases, the incoming waveform will be deteriorated, and will require processing be-

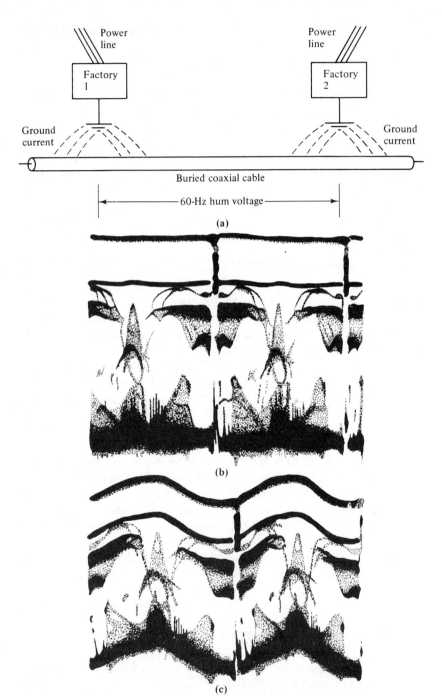

Figure 1-16 How ground currents can cause hum in a signal system. **(a)** Difference in ground potentials produces 60-Hz current flow in the coaxial cable; **(b)** video signal without hum interference; **(c)** video signal with hum interference.

fore it can be telecast with acceptable quality. Accordingly, the system designer must anticipate worst-case conditions on signal tolerances, and provide necessary facilities for optimum system operation. Moreover, receiver characteristics must be taken into account in signal processing techniques. For example, as a result of vestigial-sideband reception, an ideal step function would be reproduced with "smear." Therefore, a suitable amount of video predistortion is provided to optimize picture reproduction. Tolerances on predistortion values are comparatively tight.

The picture tube in a TV receiver has a nonlinear gamma; that is, its light output increases out of proportion to the signal voltage applied to its control grid. Accordingly, the systems designer provides facilities at the transmitter for gamma correction. In other words, a nonlinear output-input characteristic is introduced at the transmitter so that the receiver will operate with an effective gamma value of unity. Close tolerances are maintained on gamma-corrected signals, and the systems designer accordingly provides adequate test and adjustment facilities. One of the tightest signal tolerances in the color-TV system is specified for the phase of the color burst. For this reason, a vertical interval reference signal (Fig. 1-17) is provided. Its purpose is to reduce undesirable variations in hue reproduction at the color receiver. Thus, the VIR signal is a reference "burst" that assists transmitter system personnel in adjustment of various signal parameters so that all

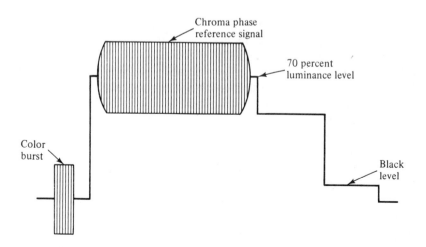

Figure 1-17 Vertical interval reference (VIR) signal is transmitted during line 20 in the vertical blanking interval of both fields.

color programs are transmitted with precise chroma phase relations.

Systems designers are also concerned with component and device tolerances. For example, cable connectors utilized in a system must mate properly with equipment connectors. Thus, the designer may specify a connector that will accommodate a radial misalignment up to 0.031 inch, or a total float of 0.062 in. This same type of connector typically accommodates 0.125-inch variation in gap between mounting surfaces. Tolerances on wire diameter are also of concern to the designer, to ensure that cables and connectors can be properly assembled in production. Tolerances on the thickness of insulation of wires are also important; out-of-tolerance insulation may prevent assembly of cables and connectors. Thus, a typical connector is rated to accept wire sizes from 14 AWG through 18 AWG, with an overall wire insulation diameter range of 0.080 to 0.130 in. To minimize the voltage standing-wave ratio (VSWR) in a system, the designer must specify a reasonably tight tolerance on cable characteristic impedance. For example, a type of cable rated for 50 ± 2 ohms Z_0 may be specified. Tolerances on microwave antenna hardware are very tight.

1-9 DESIGNING FOR SYSTEM RELIABILITY

Reliability engineering is defined as the establishment, during design, of an inherently high reliability in a product; *reliability* denotes the probability that a device will perform adequately for the length of time intended and under the operating environment encountered. However, it is not sufficient that the elements of a system have high reliability; system reliability is reduced, in the first analysis, in direct proportion to the number of key elements that it contains. In other words, if a cable system includes six amplifiers between its input terminals and its output terminals, the system reliability cannot exceed one-sixth of the reliability of an individual amplifier. Moreover, since there is an additional possibility of system failure owing to cable faults, the system reliability is lessened accordingly.

System reliability design is primarily based on *redundancy*. That is, the designer may employ several devices, each performing the same function, in order to improve the reliability of a particular function. For example, if one cable amplifier fails, provisions may be made for automatic switching of a spare amplifier into the circuit. Sometimes the input and output terminal facilities at two geographical locations may have a choice of two or more cable routes. In such a case, if the cable on one route develops a fault, the operating personnel can patch an alternate cable route into the system network. System designers provide

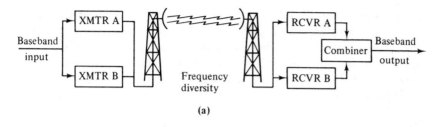

(a)

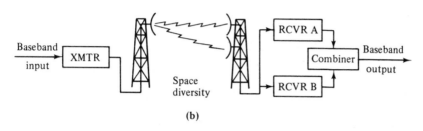

(b)

Figure 1-18   Example of diversity designs. **(a)** Frequency diversity; **(b)** space diversity.

patch panels in the more elaborate installations to maximize reliability and to minimize down time insofar as may be economically feasible. Inevitably, certain tradeoffs in reliability versus costs must be accepted.

In electromagnetic wave systems, reliability can be improved under conditions of fading by utilizing a *diversity* design. This method also has a redundant aspect, in that it can partially compensate for outages in any one of the modes. Refer to Fig. 1-18. In the *space diversity system,* the same signal is sent simultaneously over several different transmission paths, which are sufficiently separated that independent propagation conditions can be expected. On the other hand, in the *time diversity system,* the same path may be used, but the signal is transmitted more than once, at different times. Other types of diversity systems employ different frequencies or different polarizations to provide the separate transmission modes. *Diversity reception* denotes the combination and/or selection of two or more sources of received-signal energy that carry the same intelligence, but differ in strength or in signal-to-noise ratio, in order to increase reception reliability under conditions of fading. These separate sources are typically two or more antennas installed at different locations. *Diversity gain* denotes the gain in reception as a result of the use of two or more receiving antennas.

Reliability can be improved in some types of information transfer systems by the use of an *error-correcting code*. This is a code in which each acceptable expression conforms to specific rules of construction that also define one or more equivalent nonacceptable expressions, so that if certain errors occur, the result in an acceptable expression will produce one of its equivalents, and thus the error can be corrected. In a digital system, an example is *parity*. This is a method of checking the accuracy of binary numbers. An extra bit, called a parity bit, is added to a number. If even parity is used, the sum of all 1's in the number and its corresponding parity bit is always even. If odd parity is used, the sum of the 1's and the parity bit is always odd.

1-10   SYSTEM DESIGN FOR MAXIMUM INFORMATION TRANSFER

Since information (intelligence) is not entirely predictable, and may be completely unpredictable, it is not possible to transmit an unlimited amount of information through a limited channel over a finite period of time. An important aspect of system design is to realize the maximum information capability of a subsystem or of a channel. Engineers employ the *bit* unit in quantitative measurement of information transfer. This is an abbreviation of *binary digit,* and the bit unit is the most basic element of information. If the system under consideration is ideal, a bit may be regarded as a *choice* between two events that are *equally probable*. For example, a signal may be "on" or it may be "off" at a particular instant. In terms of binary arithmetic, the "off" state represents the digit 0, and the "on" state represents the digit 1. Thus, 0 and 1 are *information bits*. If a line is resting at 0, and suddenly changes to 1, one bit of information has been transferred. Then, if the line level suddenly changes to 0 at a following instant, two bits of information have been transferred.

It is evident that a digital pulse is a bit, and that a pulse can convey various kinds of information, depending upon the circuit that it energizes. Refer to Fig. 1-19. At the origin, information transfer can take place to one of four output terminals. These four choices are denoted *C, D, E,* and *F.* Two bits of information will select any one out of the four eventualities. For example, if terminal *E* is to be energized, the first pulse is applied to a logic circuit that represents "If *A,* then not *B,*" or, conversely, "If *B,* then not *A.*" This circuit is energized to transfer a pulse to *B.* From *B,* the second bit of information is applied to a logic circuit that represents "If *F,* then not *E,*" or, conversely, "If *E,* then not *F.*" This circuit is energized to transfer a pulse

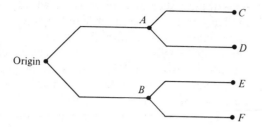

Figure 1-19   One out of four choices may be selected in two steps.

to *E*. Thus, two bits of information have served to select one out of four choices.

Next, refer to Fig. 1-20. This diagram illustrates the fact that three steps are involved to select one out of eight choices. In other words, to select *J*, the first bit will represent "*A* and not *B*"; the second bit will represent "*D* and not *C*"; the third bit will represent "*J* and not *I*." Thus, a definitive equation is written:

Number of bits $= \log_2 N$

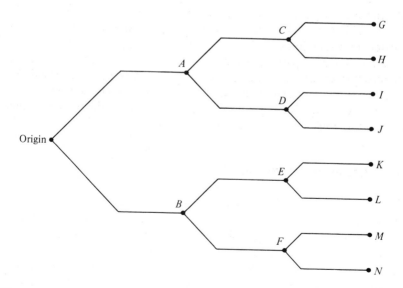

Figure 1-20   One out of eight choices may be selected in three steps.

where N represents the number of choices from which selection is to be made, and $\log_2$ denotes the logarithm of N to the base 2.

System design engineers recognize that five bits are required to identify any letter in the English alphabet by means of digital pulses. If there are five letters to a word, and 300 words on the page of a book, it requires approximately 10,000 bits to transmit a printed page from a book through a communication system in the form of digital pulses. In other words, the amount of information contained on a printed page is equal to 10,000 bits. Consider next the number of bits that are required to transfer the information on a printed page via television. It can be shown that the picture-tube screen must reproduce at least 250,000 spots of light to display the information intelligibly. A rapid reader can scan a printed page for its chief information in 10 sec. Accordingly, the televised image must be repeated at least 300 times. This message requires $300 \times 250,000$, or 75×10^6 bits, instead of the 10,000 bits required for digital pulse transmission. If a storage-type picture tube is employed, only 250,000 bits will be required. However, 25 times as many bits will be required for the TV transmission as for the digital pulse transmission.

System designers increase information transfer in communication channels by various *multiplexing* techniques. This is a process for combining more than one signal for transmission over the same path, or within the same channel. There are two widely used methods of multiplexing. *Time-division multiplexing* employs the principle of time sharing among signal sources. *Frequency-sharing multiplexing,* on the other hand, also called frequency-division multiplexing, employs subcarriers on which individual signal sources are modulated. For example, time-division multiplexing is extensively utilized in telemetry systems; frequency-sharing multiplexing is employed in stereophonic broadcasting, and in color-television broadcasting. Both time-division multiplexing and frequency-sharing multiplexing techniques are being investigated for quadraphonic broadcasting at this writing.

1-11　TELEPHONE SYSTEM BASICS

Almost everyone is familiar with the sound-powered telephone circuit shown in Fig. 1-21(a). A telephone receiver is connected at each end of a line, to serve alternately as a receiver or as a transmitter (microphone). A more efficient arrangement employs a separate carbon transmitter, as depicted in Fig. 1-21(b). A carbon transmitter is a simple form of audio amplifier. In turn, voice currents can be carried over lines to greater distances. Improved operation is provided by the

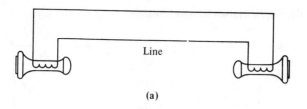

(a)

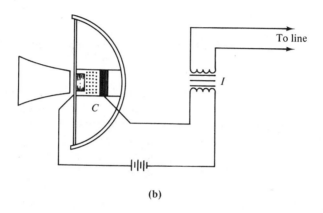

(b)

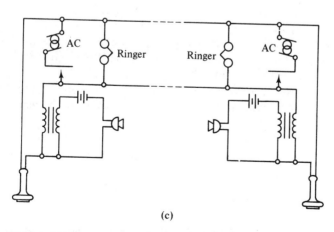

(c)

Figure 1-21 Fundamental telephone arrangements. **(a)** Sound-powered telephone circuit; **(b)** carbon transmitter (microphone) circuit; **(c)** elementary telephone circuit with ringer facilities.

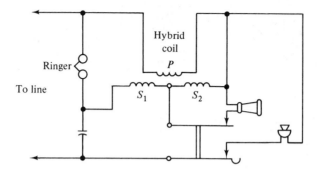

Figure 1-22   An anti-sidetone station circuit.

induction coil I in the diagram. An induction coil is essentially an impedance-matching transformer that matches the low impedance of the carbon microphone to the higher impedance of the line. An elementary telephone circuit with ringer facilities is depicted in Fig. 1-21(c). A 20-Hz AC source is provided for energizing the two ringers (bells) on the line. Because of the shunt signal paths in the configuration, efficiency is less than could be realized. However, this is a *tradeoff* that is accepted by the designer in order to minimize the switching circuitry that is utilized.

One of the undesirable features of the simple arrangement depicted in Fig. 1-21 is the presence of *sidetone* in the receiver at the transmitting location. Sidetone is defined as the reproduction, in a telephone receiver, of the sounds produced by the transmitter of the same telephone set; that is, a loud reproduction of one's own voice in the receiver of a telephone set when one is speaking into the mouthpiece. Therefore, system designers employ anti-sidetone circuitry with a *hybrid coil,* as shown in Fig. 1-22. A hybrid coil, also called a *bridge transformer,* has effectively three windings. It is connected into the four branches of the circuit so that voice currents from the transmitter flow into the line, but are cancelled in the receiver of the same telephone set. This advantage is obtained at the expense of a tradeoff in efficiency, because half of the voice current from the transmitter is dissipated in the hybrid-coil configuration.

Phantom Circuits

If two pairs of wires of the proper type are available, a third transmission path may be obtained in a system by using one pair of wires for one side of the third circuit, and the second pair for the other side.

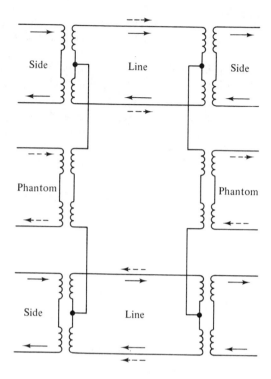

Figure 1-23 Basic phantom circuit.

This basic arrangement is shown in Fig. 1-23. Repeating coils are used in a bridge configuration. For satisfactory phantom operation on open-wire lines, it is necessary that the two wires of a pair have approximately equal resistances and that the lines be suitably *transposed* to prevent objectionable *crosstalk* among the three constituent circuits of each phantom group and between nearby phantom groups in the system.

In cable circuits, crosstalk considerations require the use of *quadded* conductors, suitably spliced at suitable intervals to minimize side-to-side and phantom-to-side crosstalk. A quadded cable is defined as a cable in which some or all of the conductors are in the form of quads, consisting of four separately insulated conductors twisted together. A *spiral-four quad* consists of four wires laid together and twisted as a group, the diagonally opposite wires being used as a pair. Domestic lead-covered quadded cables are of multiple-twin design, consisting of two twisted pairs, with the pairs twisted together. The phantom-deriving repeating coils must be well balanced. Phantom circuits tend to be noisier than their side circuits, or nonphantomed circuits, particularly if the lines or equipment are not maintained in good electrical balance.

Cables and lead-covered paper-insulated cables may be operated with or without *loading*. Loading is the addition of series inductance at regular intervals along a line. This added inductance increases the impedance of the circuit and decreases the series losses owing to conductor resistance. Loading increases any shunt losses caused by leakage and also causes the line to have a *cutoff frequency* above which the line loss becomes very high. The loss is actually increased by loading at frequencies above about 90 percent of the cutoff frequency. The approximate cutoff frequency and the nominal impedance of a loaded line are given by the following equations:

$$f_c = \frac{1}{\pi\sqrt{LC}}$$

$$Z = \sqrt{\frac{L}{C}}$$

where f_c = cutoff frequency in hertz.
Z = nominal impedance (resistance) in ohms.
L = inductance of a loading coil in henrys.
C = capacitance of a loading section in farads.

The usual method of designing loading systems is by first indicating the spacing of the loading coils in feet, then the inductance of the side circuit in millihenrys (mH), and last the inductance of the phantom circuit (if there is one) in mH. For example, 6000-88-50 represents a loading system in which the loading coils are spaced 6000 feet apart, the inductance of the side-circuit loading coils is 88 mH, and the phantom circuit employs loading coils with 50-mH inductance. Again, 6000-88 denotes a nonphantomed circuit loaded with 88-mH loading coils spaced 6000 feet apart. A speech-frequency circuit could consist of 88-mH loading coils spaced at one-mile intervals. However, a carrier-frequency circuit requires a higher cutoff frequency; it could consist of 6-mH loading coils spaced at 1/4-mile intervals. The extent of the system designer's problem becomes apparent when it is recognized that the capacitance of a cable circuit between San Francisco and New York is approximately 180 μF.

The range of voice-frequency circuits can be extended by the use of line amplifiers called *repeaters*. Three basic types are termed the 21-type, the 22-type, and the 4-wire repeater. A 21-type repeater, depicted

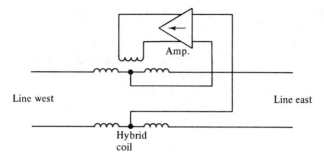

Figure 1-24   Arrangement of a 21-type repeater.

in Fig. 1-24, has a circuit arrangement that requires no balancing networks, and stability (freedom from singing) is realized by the balance between the lines on the two sides of the repeater. Since freedom from self-oscillation (singing) depends upon equality of impedances on the two sides of the repeater, this configuration is not suitable for use on circuits that are made up from more than one kind of facility. Thus, the best location for a 21-type repeater is at the midpoint of a line or circuit. It is possible to use a 21-type repeater on a loaded circuit, but the usable gain is subject to wide variations, and in some cases it may be very small. Type-21 repeaters can be operated in tandem, but in general there is little transmission advantage in this expedient.

A 22-type repeater, depicted in Fig. 1-25, employs a circuit arrangement with two balancing networks. Stability is obtained by the balance between the impedance of each network and its associated line.

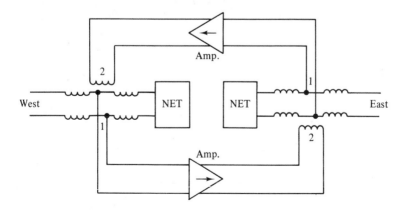

Figure 1-25   Arrangement of a 22-type repeater.

This type of repeater may be used at a circuit terminal or at intermediate points and will operate satisfactorily in tandem with other 22-type repeaters. A 22-type repeater can be used on any type of stabilized line for which suitable balancing networks are available. These repeaters may also be used on nonstabilized lines, but at considerably reduced gain. The maximum usable gain (MUG) of a 22-type repeater is limited to a value that provides adequate margin against singing or near-singing. In some systems, crosstalk or echoes, rather than singing, may limit the usable gain.

A four-wire repeater (Fig. 1-26) consists of two one-way amplifiers that operate in opposite directions. Each amplifier processes signals in one direction only. Four-wire repeaters can be used at circuit terminals and can be used in tandem with other four-wire repeaters. It is an inherently stable arrangement and can provide substantial gain on a line that has irregular impedance. Usable gain is generally limited by crosstalk, noise, or transmission variations. No balance requirements are involved. On the other hand, a 21-type or a 22-type of repeater requires careful balance. Refer to Fig. 1-25. If the two amplifiers were merely connected together, an oscillator configuration would result, and the arrangement would sing or howl. This difficulty is avoided by the use of a bridge circuit arrangement that effectively isolates the two amplifiers.

The action of a hybrid coil is similar to that of a Wheatstone bridge. When a voltage is applied across a particular pair of terminals of the bridge, no current will flow through an impedance connected across the other two terminals of the bridge, if balance conditions regarding the ratios of the impedances of the four arms of the bridge are observed. Thus, if a voltage is impressed on terminal 2 in Fig. 1-25, no current will flow in terminal 1 if the impedances of the line and network are identical. The greater the difference between the imped-

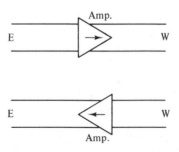

Figure 1-26 Four-wire repeater arrangement.

ances of line and network, the greater will be the amount of current that flows in terminal 1, and the greater will be the tendency of the repeater to sing. Since perfection cannot be attained even in the most carefully designed telephone circuits, there is always some transmission across the hybrid coil. Consequently, the amount of amplification that can be provided by a repeater without singing is limited. The sum of the gains of the two amplifiers must always be greater than the sum of the losses across the two hybrid coils in order to prevent singing. In practice, *to allow for variations in the circuit, the designer specifies a margin between total loss and total gain.*

The *return loss* between two impedances is a measure of the similarity between the impedances. These might be the line and network impedances, or the impedances of two types of line. It is expressed in dB, and is equal to 20 times the logarithm of the reciprocal of the numerical value of the reflection coefficient; viz., return loss:

$$R = 20 \log \left[\frac{Z_1 + Z_2}{Z_1 - Z_2} \right]$$

The loss across a conventional hybrid coil is about 6 dB greater than the return loss. Return loss is defined as the difference between the power incident upon, and the power reflected from, a discontinuity in a transmission system, or as the ratio in dB of these power values. Since speech transmission through a repeater must be practically uniform over the band of frequencies that is employed (approximately 200 to 2800 Hz), it is evident that transmission loss across a hybrid coil must be sufficiently great at all frequencies in this band to prevent singing at any one frequency. *System designers generally utilize filters to prevent singing at frequencies outside of the speech-transmission band.*

It is evident that good balance cannot ordinarily be obtained by matching the line impedance at a single frequency by means of a simple network consisting of a resistor and a capacitor or an inductor. Excellent balance between line and network could, of course, be obtained by duplicating in the network each element of the line. However, this is an impractical solution. Therefore, fixed networks are utilized by the designer, or variable networks are provided that can be adjusted in the field to approximately match the characteristic impedance of the types of lines in general use. These networks are furnished as a part of the repeater.

To obtain good balance between a line and a network, the line must be reasonably uniform. Where apparatus is placed between the hybrid coil and the line, duplicate apparatus is usually placed between the hybrid coil and the network. A line irregularity that is distant from the repeater is less important than one that is close to the repeater. If a repeater section is composed of two dissimilar lengths of line in tandem, L_1 and L_2 with characteristic impedances Z_1 and Z_2, and if the repeater connected to L_1 matches Z_1, and the repeater connected to L_2 matches Z_2, then the return loss at the junction of L_1 and L_2 is R, as explained above, but the return loss at the repeater connected to L_1 is equal to $R + 2A$, where A is the attenuation of L_1. This is apparent by noting that a current starting at the repeater would traverse L_1, be partially reflected at the junction of L_1 and L_2, and traverse L_1 again before the reflected voltage reached the repeater.

Carrier Telephony

Carrier telephony is defined as a system of telephone communication wherein several voice or coded signals may be transported simultaneously over a single circuit. Modulation and demodulation processes are utilized to shift the band of voice or code frequencies to higher positions in the frequency spectrum. *Carrier telephony is used by system designers to increase the signal-handling capability of a wire line or cable network.* Modern carrier-telephony technology employs single-sideband transmission. In other words, the carrier and one of the sidebands is suppressed in the amplitude-modulated signal. The essential features of a modern carrier telephony system are shown in Fig. 1-27. Proceeding from left to right in the diagram, observe the following features:

First, generation of a carrier wave is provided by a carrier oscillator. This carrier voltage has a typical frequency of 25 kHz. Second, modulation of the carrier is generally accomplished with an amplitude modulator. Third, it is also possible to frequency-modulate or to phase-modulate the carrier. Frequency shift keying (FSK) modulation, a specialized variety of frequency modulation, is also employed. Fourth, carrier suppression was accomplished by use of filter networks in the older carrier telephony systems. Today, modulators that automatically eliminate the carrier from the modulator output signal are utilized; these are called *balanced modulators.* In the example of Fig. 1-27, the lower sideband and carrier have been suppressed by a filter; only the upper sideband is being transmitted. Fifth, as previously noted, the transported signals must be amplified at regular intervals by means of repeaters to avoid "losing" the signals in the prevailing noise level. Sixth, at the receiving end of the line or cable, the missing carrier must be reinserted at the demodulator in order to reproduce the original voice or code

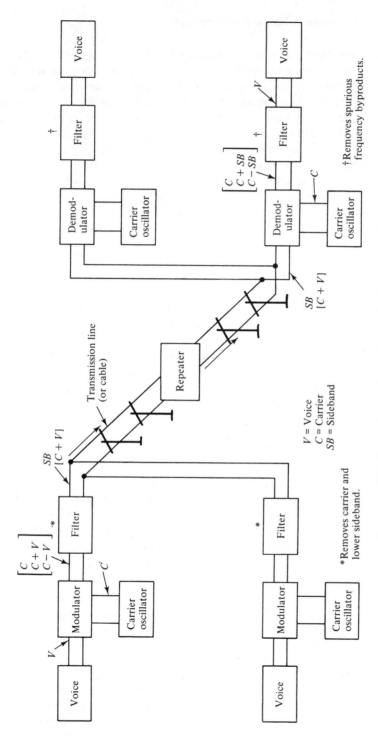

Figure 1-27 Typical two-channel one-way carrier system.

†Removes spurious frequency byproducts.

*Removes carrier and lower sideband.

V = Voice
C = Carrier
SB = Sideband

39

signal waveform. Spurious frequency byproducts are removed by appropriate filters. The modulation and demodulation processes are generally accomplished with copper-oxide or germanium devices called *varistors*. Seventh, multichannel operation is commonplace, utilizing the technique that has been described, with different carrier frequencies.

System designers must cope with practical problems in carrier telephony such as maintenance of a minimum distortion level, prevention of crosstalk in multichannel communication, and repeater design that effectively directs signals into one line or cable, but not into another. An open-wire transmission line has a typical cutoff frequency of 150 kHz. A telephone signal has an upper frequency limit of approximately 2.8 kHz. Multichannel carrier telephony systems may utilize the frequency spectrum from 2.8 kHz to 150 kHz on a typical open-wire transmission line. To efficiently utilize the available frequency spectrum of a transmission line or cable, audio channels are reallocated in carrier systems. This reallocation is termed *frequency-division multiplexing.* One audio channel may occupy a 2600-Hz "spread" in a 2800-Hz band. Another audio channel may occupy a 2600-Hz "spread" in a band from 3100 to 5700 Hz. In other words, a guard band or gap is provided between the reallocated channels to avoid interference owing to residual overlapping. A skeleton block diagram for a basic carrier system is shown in Fig. 1-28. Each terminal includes various channels, and each channel provides for an individual two-way voice communication.

Consider the path for voice-signal transmission in the example of Fig. 1-28. Starting from the west terminal, voice signals V are fed to the modulator from an adjacent switchboard and the carrier current C enters the modulator from an associated modulator. A bandpass filter is provided to pass only the desired range of frequencies; it is essentially an LC filter configuration. Next, at the east terminal, the line applies the signals to a common receiving amplifier. Other bandpass filters are connected at the amplifier output to select desired ranges of frequencies. A selected frequency range is applied to a demodulator section with the output from an oscillator that inserts the missing carrier frequency. In turn, the sideband frequencies with the inserted carrier frequencies are processed through the demodulator. The demodulator output contains various frequency components; one of these components is the original voice frequencies V.

A subsequent low-pass filter passes the voice frequencies V and feeds them to a receiving switchboard. This low-pass filter has typical frequency limits from nearly zero to 2800 Hz. Observe that the same process that has been described is also used for transmission in the opposite direction from east to west. *In practice, the modulator and*

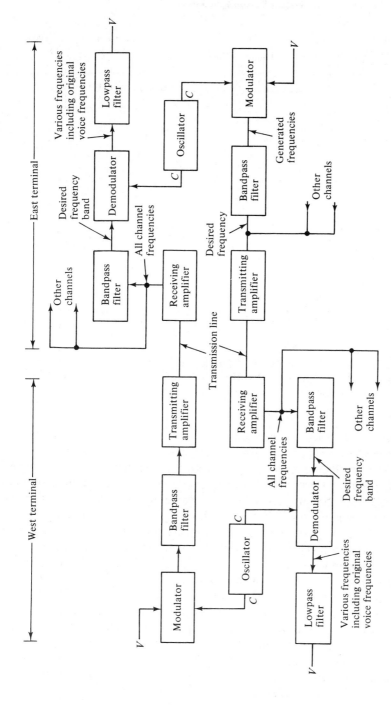

Figure 1-28 Skeleton block diagram of a basic carrier system.

41

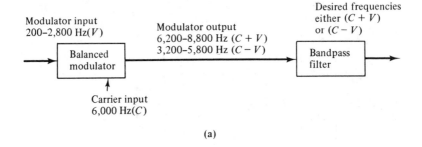

(a)

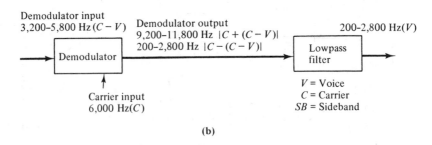

(b)

Figure 1-29 Frequency relations for a typical carrier system. **(a)** Relations in modulation and transmission of one sideband; **(b)** relations in demodulation and transmission of voice signal.

demodulator in the transmitting and receiving sections of a carrier channel generally obtain their carrier frequency from a common source, as shown. The modulation and demodulation functions in a carrier channel are termed a *modem*. Frequency relations for modulation and transmission of one sideband, and demodulation and transmission of a voice signal are shown in Fig. 1-29.

Modems in modern carrier telephony systems often utilize copper-oxide varistors. These are essentially rectifiers. The arrangement of a copper-oxide varistor is depicted in Fig. 1-30. It provides design advantages of simplicity, stability, and very long life. A typical varistor employs four copper-oxide discs with a diameter of 3/16 in. These discs are assembled with suitable connections in a sealed housing. Each of the discs is a perforated circular copper plate coated with a layer of copper oxide on one side. The copper surface is connected to a wire, and the oxide surface is contacted by a special vaporized-metal conductor. The voltage-resistance characteristic of a typical varistor is shown in Fig. 1-31. *Designers also employ germanium diodes and silicon diodes as varistors.*

A balanced-bridge suppressed-carrier modulator using varistors is

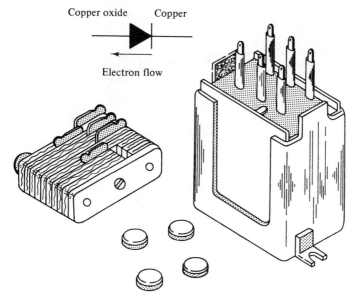

Copper oxide   Copper

Electron flow

Three stages of assembly

Figure 1-30   Arrangement of a copper-oxide varistor.

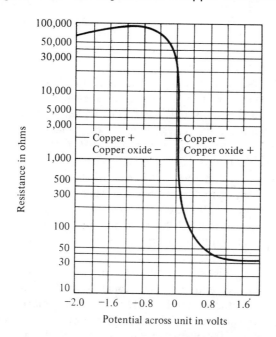

Copper +
Copper oxide −

Copper −
Copper oxide +

Resistance in ohms

Potential across unit in volts

Figure 1-31   Voltage-resistance characteristic of a typical varistor.

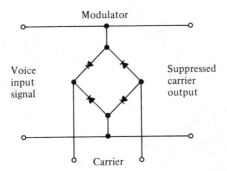

Figure 1-32   Suppressed-carrier modulator arrangement.

shown in Fig. 1-32. In this arrangement, the voice signal is fed into the leads that go to the output. The varistor Wheatstone bridge places a resistance across the leads that varies in accordance with the instantaneous amplitude of the carrier voltage. Since the voice signal does not have a constant voltage source, this varying resistance controls the output signal level. Because the arms of the bridge are nonlinear resistances, the voice signal is amplitude-modulated on the carrier frequency. The carrier voltage does not feed through into the output, inasmuch as the bridge is balanced. On the other hand, the input voice signal does feed through to the output in this type of balanced modulator. Therefore, the feed-through voice-frequency signal is subsequently removed by means of a high-pass filter.

Consider the operation of a balanced modulator as a switching bridge; each varistor can be regarded as a carrier-controlled switch. With reference to Fig. 1-33, only the carrier voltage has been applied in (a). R_1 represents the input resistance of the voice-signal source; R_2 represents the resistance of the modulator load. Observe the indicated carrier polarity. Terminal E is positive, and terminal G is negative at this time. Under this condition, no current flows through the varistors, which are effectively open circuits, as depicted in (b).

During the next half-cycle of the carrier voltage, terminal G becomes positive, and terminal E becomes negative. In turn, the varistors conduct current and shunt a comparatively low resistance across the input-output leads. In the first analysis, the varistors can be regarded as short circuits at this time, as depicted in (d). Refer to Fig. 1-34. Here, both the carrier and the voice signal sources are applied to the balanced bridge arrangement. In (a), the carrier polarity makes terminal G negative and terminal E positive. Consequently, the varistors are open circuits. If the instantaneous polarity of the voice signal is

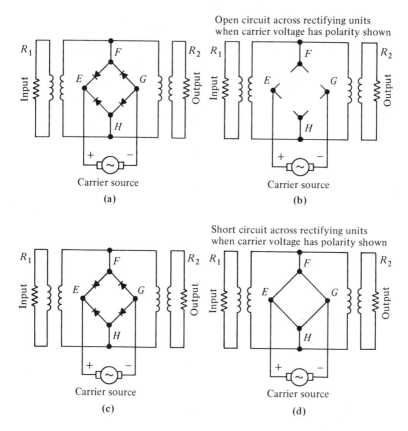

Figure 1-33   Operation of a balanced modulator as a switching bridge.

such that the voice current flows down through R_1, and upward through R_2 via transformer action, there is zero current flow through the varistors. Note that R_1 and R_2 are equal in value; the input and output circuits have matched impedances.

At another instant, the carrier voltage will have the same polarity as above, but the voice-signal polarity will have reversed. Although the varistors are still open circuits, the voice currents now flow up through R_1 and down through R_2. In other words, there is then a reversal in direction of voice-current flow through R_2. Next, when the carrier voltage reverses its polarity, as depicted in (b), there is zero voice-current flow through R_2. At this time, terminal G is positive and terminal E is negative. Accordingly, the varistors are short circuits. This short circuit across the primary of the output transformer prevents any

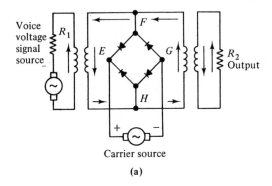

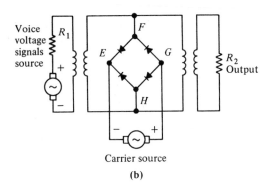

Figure 1-34 Balanced modulator operation with carrier and voice signal voltages.

transfer of energy to the secondary, regardless of the voice-signal polarity. Assume next that the carrier-voltage polarity that permits voice-current flow through R_2 is positive, and that the reverse carrier-voltage polarity is negative. With reference to (a), observe that one voice cycle of voltage and many carrier cycles of voltage are on the same time axis, but are not combined in the modulator. The carrier-voltage polarity is indicated by positive and negative signs. During positive half-cycles of the carrier, voice current flows through R_2. The amplitude of voice current depends on the instantaneous value of the voice voltage, because the varistor resistance depends on the total voltage applied across it (carrier and voice voltages).

Refer to Fig. 1-35. The carrier amplitude is greater than the voice-signal amplitude (actually much greater than shown in the diagram). This disproportion of voltages optimizes modulator operation. Note in

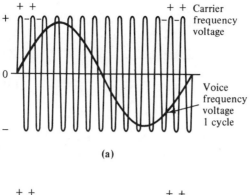

(a)

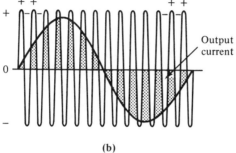

(b)

Figure 1-35   Output waveform development by a balanced bridge modulator.

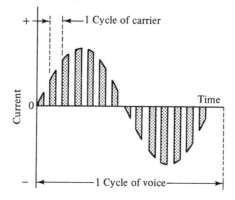

Figure 1-36   Output waveform from the balanced bridge modulator.

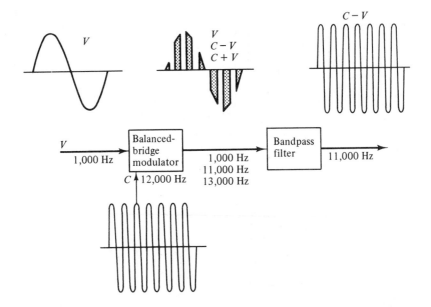

Figure 1-37   Basic frequency relations in the modulator and bandpass filter section.

(b) that the shaded area denotes the amount and polarity of voice current through R_2 in Fig. 1-34. Next, the signal waveform developed in the modulator output is depicted in Fig. 1-36. Although this waveform has many frequency components, only a few are utilized by the carrier system. Components include the original voice frequency, an upper sideband, a lower sideband, and various harmonics. A skeleton block diagram of a balanced bridge modulator followed by a bandpass filter is shown in Fig. 1-37. It indicates the location of several frequencies at the transmitting terminal of a simple carrier system. Observe that the voice frequency V (1kHz), and the carrier frequency C (12 kHz), are combined in the modulator to develop the output waveform. In turn, this output waveform includes the frequencies V, $(C - V)$, $(C + V)$, and an array of harmonics. V has a frequency of 1 kHz, $(C - V)$ has a frequency of 11 kHz, and $(C + V)$ has a frequency of 13 kHz. However, the bandpass filter permits the passage of the 11-kHz component only. In other words, only the lower sideband appears at the bandpass filter output.

2

Basic Radio
Network Design

2-1 GENERAL CONSIDERATIONS

A standard radio broadcast network has the general organization shown
in Chart 2-1. Its subsystems must necessarily work together as a team.
System design has evolved over the years from comparatively simple
installations, and its implementation becomes more sophisticated with
the rapid progress of electronic technology. An instructive point of
entry concerns broadcast channel allocations. An *allocated channel* is a
frequency interval in the radio-frequency spectrum that is assigned to a
specific user, and which is designated for a particular type of use by a
regulatory agency such as the Federal Communications Commission
(FCC) in the United States of America. Standard amplitude-modulated
(AM) broadcast channels are allocated at 10-kHz intervals, as exemplified in Fig. 2-1. In other words, the channel bandwidth is 10 kHz, and
the nominal audio-frequency range for a channel is 5 kHz.

Since AM broadcast stations are so closely spaced in this *broadcast band,* good practice in system design requires that RF carrier-
frequency drift be minimized. Accordingly, the FCC stipulates that
each station carrier frequency must be maintained within ±20 Hz of
its assigned value. The standard broadcast band extends from 535 to
1605 kHz; it can (in theory) accommodate a maximum number of 107
transmitting stations in one *service area.* A service area is the geographical region surrounding a broadcasting station in which the field
strength of its radiated signal is sufficient for satisfactory reception.
Interference among various broadcasting stations is a basic problem

Chart 2-1 Standard radio broadcast system.

STANDARD RADIO BROADCAST SYSTEM

- Transmitter installation
  - Studios
  - Control booths
  - Studio-transmitter links
  - Transmitting equipment
  - Transmitting antenna(s)
- Microwave receivers
  - Remote-relay
  - Remote-control
  - Remote-pickup
- Network facilities
  - Telco lines
  - Nemo lines
  - Local subsystem lines
- Field facilities
  - Remote-pickup and base-station equipment
  - Portable transmitters
- Recording facilities

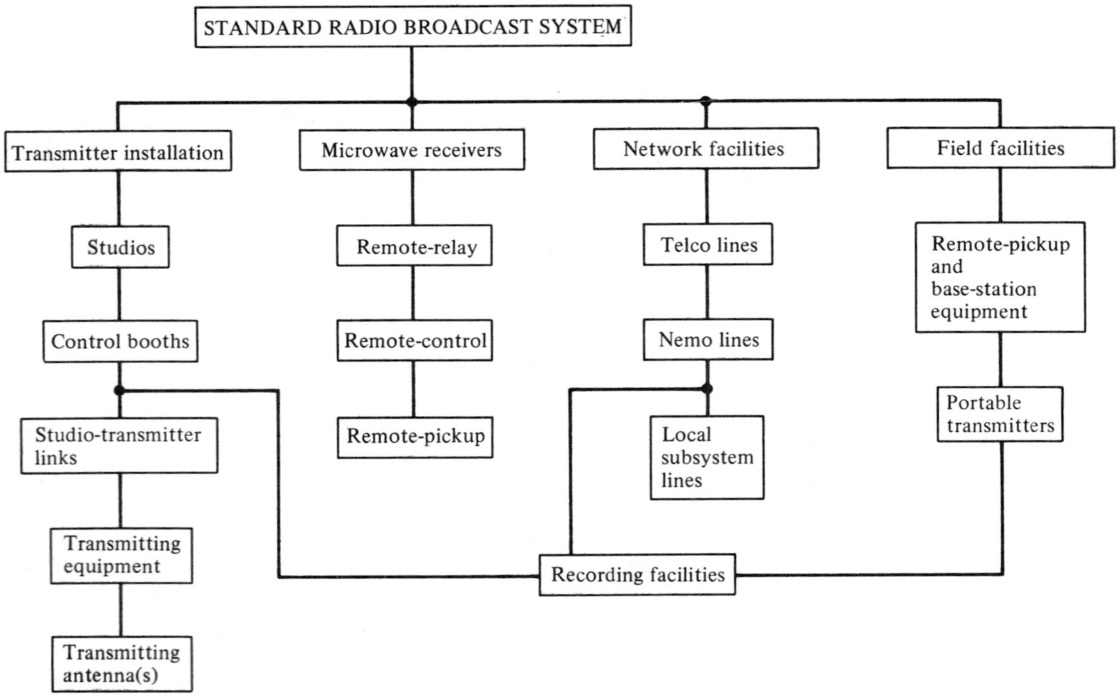

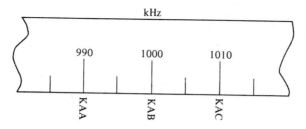

Figure 2-1 Example of standard AM broadcast channel allocations.

in network operation. Accordingly, stations that operate on the same channel, or on adjacent channels, are geographically separated. System designers also control interference by means of directional transmitting antennas.

Service areas are subdivided into *primary service areas* and *secondary service areas*. A primary service area, also termed a primary area, is the region in which reception is not normally subject to noticeable interference or fading. On the other hand, the secondary service area, also called a secondary area, is the region within which satisfactory reception can be obtained only under favorable conditions. During the daytime, reception in the standard broadcast band is obtained only by means of the ground wave. In the first analysis, the intensity of the ground wave may be regarded as inversely proportional to the distance from the transmitting antenna, as depicted in Fig. 2-2. The ground wave has a typical usable range of 40 miles, although a very high power station transmitting over terrain of comparatively high conductivity may have ground-wave radiation usable up to several hundred miles.

It is of interest to note that the ground wave comprises three different types of waves, termed (1) the *surface wave,* (2) the *direct*

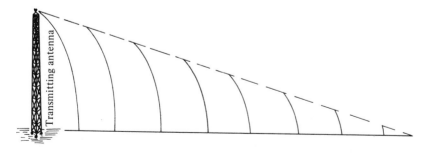

Figure 2-2 Generalized representation of ground-wave intensity.

wave, and (3) the *ground-reflected wave.* A surface wave travels in contact with the ground and terminates on the earth's surface. It is the radiation component that is normally received from a standard broadcast station. The direct wave is a line-of-sight component that is propagated directly through space. It does not touch the ground; a direct wave cannot be received unless the transmitting antenna is unobstructed and visible from the receiving antenna. Note that a receiving antenna may intercept both a surface wave and a direct wave if suitably located. A direct wave will sometimes strike the ground and become reflected in a manner such that it is intercepted by a receiving antenna. This radiation component is called the ground-reflected wave. It may combine with the direct wave to energize a suitably located receiving antenna; in such a case it is termed the *space wave,* or the *resultant wave.* A space wave is sometimes called a *ground wave.*

An *intermittent service area* denotes a limited region beyond the primary service area wherein the ground wave can be received, although reception is marginal owing to fading and/or interference. The secondary service area is useful only after sunset, when sky-wave radiation is reflected back to the earth's surface, as depicted in Fig. 2-3. A sky wave is also called an *indirect wave.* When the ionosphere is at a suitable height, and a direct wave strikes the ionosphere at a suitable angle, it is reflected and/or refracted as an indirect wave into the secondary service area of the transmitter. Since the ionosphere tends to change in height and density from time to time, indirect-wave reception is subject to more or less fading. This term denotes a variation in signal field strength. It can be compensated to some extent by automatic volume control (AVC) action at the receiver. However, in many

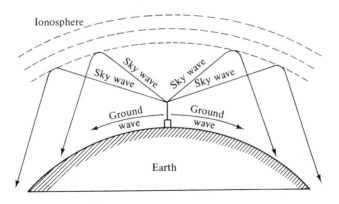

Figure 2-3   Sky waves are reflected back to earth after sunset.

situations, an indirect wave will fade to inaudibility, only to return to a comparatively high level after a period of time.

A comparison of the primary and secondary service areas for a network of three standard broadcast stations is pictured in Fig. 2-4. Observe that although their primary service areas do not overlap, their secondary service areas occupy essentially the same geographic regions. Within the intermittent service areas just beyond the primary service areas, one station may interfere with itself owing to the reception of both a sky wave and a ground wave at the receiving site, as shown in Fig. 2-5(a). In other words, *the ground wave and the sky wave do not have a fixed phase relation,* and the phase of the sky wave varies from time to time. In turn, the ground wave periodically reinforces and cancels the sky wave at the receiving antenna. Moreover, within the secondary service area, one station may interfere with itself owing to

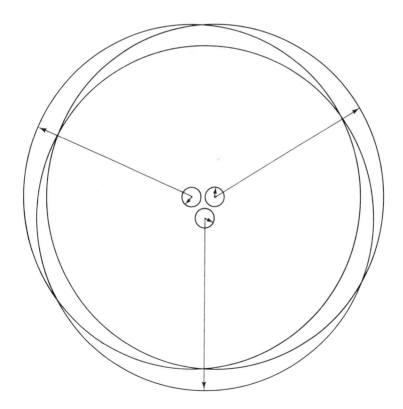

Figure 2-4   Comparison of primary and secondary service areas for a network of three standard broadcast stations.

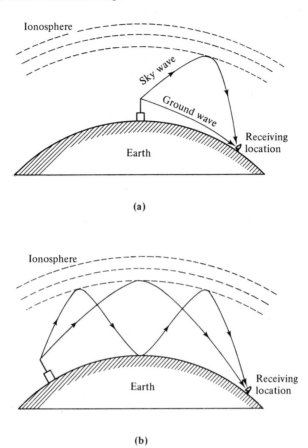

(a)

(b)

Figure 2-5 A broadcast station can interfere with itself. **(a)** Interference between the ground wave and the sky wave; **(b)** interference between two of the sky waves.

the reception of two sky waves at the receiving site, as depicted in Fig. 2-5(b). That is, the height of the reflecting layer in the ionosphere is not fixed, and *one sky wave periodically reinforces and cancels the second sky wave at the receiving antenna.*

2-2   NETWORK CHANNELS AND STATION POWER

System designers exert basic control over interference between stations by means of channel specifications, geographical allocations, and radiated power values. Basic classes of channels include:

1. *Clear Channel.* This is a channel that serves an extensive area. It may be allocated to one high-power station, or to more than one station operating at a lower radiated power level. Directional antennas are sometimes required to avoid interference between clear-channel stations. System design is directed to minimizing interference over the secondary service area, and to elimination of interference over the primary service area.

2. *Regional Channel.* This is a type of standard broadcast channel that accommodates several stations, each with a maximum power of five kilowatts. Their primary service areas may be limited as required to avoid objectionable interference.

3. *Local Channel.* This is another type of standard broadcast channel in which several stations are accommodated, each of which has a maximum power of 1 kW during the day, and 250 watts at night. Their primary service areas may be limited as required to avoid objectionable interference.

Co-channel interference denotes interference between two signals of the same type from transmitters operating in the same channel. This type of interference can often be minimized by suitable design of antenna systems. As an illustration, a basic example of directional radiation is shown in Fig. 2-6. Directivity results from localized reinforcement or cancellation of the radiated waves from the two antennas. Interference is also minimized by allocating particular types of channels to certain classes of stations. Basic classes of stations include:

1. *Class I Stations.* These are powerful stations that operate on clear channels and which have extended primary and secondary coverage. A Class I station operates at a power level in the range from 10 to 50 kW.

2. *Class II Stations.* These are also clear-channel stations; their

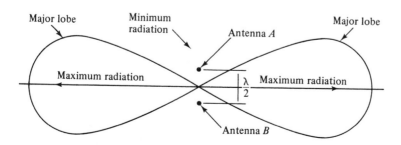

Figure 2-6 Horizontal radiation pattern of two vertical antennas spaced a half-wave apart and fed in phase.

primary service areas are limited by potential interference from Class I stations and from other Class II stations. Directional antennas are often required to avoid interference. Class II stations are categorized into three groups. A *Class IIA station* is permitted to broadcast at all times and must operate with a power level of at least 10 kW at night, but not more than 50 kW at any time. A *Class IIB station* is also permitted to broadcast at all times, and must operate at a power level of at least 250 watts, but not more than 50 kW. A *Class IID station* is permitted to operate only during the daytime, or during some limited time period. It must operate at a power level of at least 250 watts, but not more than 50 kW.

3. *Class III Stations.* These are stations that operate on regional channels with other stations, at a power level not greater than 5 kW. Directional antennas may be required to avoid interference. Class III stations are categorized into two groups. Thus, a Class IIIA station operates at a power level of at least 1 kW, but not more than 5 kW. Its service area may be subject to interference. A Class IIIB station operates at a power of at least 500 watts but not more than 1 kW at night, and at 5 kW during the day. Its service area may be subject to interference.

4. *Class IV Stations.* These are stations that operate on a local channel and that are designed to provide service primarily to a city or town and the neighboring suburban and rural areas. A Class IV station operates at a power level of at least 250 watts. It must not operate at more than 1 kW during the day, nor at more than 250 watts at night. Its service area may be subject to interference.

System designers originally established 10-kHz channels for radio broadcast networks because opinion was virtually unanimous that an audio range to 5 kHz was entirely adequate for program reproduction. By way of comparison, telephone system designers employ an audio-frequency range from 300 to 3000 Hz for commercial voice communications. This comparatively limited frequency range is quite adequate for *intelligibility,* although "naturalness" of reproduced vocal tones is impaired to some extent. An audio-frequency range from 100 Hz to 5 kHz greatly improves the quality of reproduced vocal tones. Even in this present-day environment of advanced technology, the vast majority of AM radio listeners are fully satisfied with an audio-frequency range up to 5 kHz. In turn, there has been no stress placed on increase of channel bandwidths in the AM radio system.

However, there is a minority of the radio audience that desires high-fidelity reproduction, with an audio-frequency range from 20 Hz to 20 kHz. This service is provided by frequency-modulation (FM) broadcast stations. Because of this trend to hi-fi reproduction, system designers have established wide-band AM transmission in selected locations that are free from adjacent-channel interference. These wide-band AM transmitters operate in a channel that extends from 30 Hz to 12 kHz. It is of interest to note that most AM radio receivers cannot reproduce this frequency range.

2-3 FM SYSTEM DESIGN CONSIDERATIONS

With the trend to high-fidelity reproduction, system designers supplemented the standard AM broadcast networks with establishment of 100 FM broadcast channels. These allocations comprise a spectrum from 88.1 to 107.9 MHz. Each FM channel has a bandwidth of 200 kHz. System designers have established three classes of FM stations, as follows:

1. *Class A Stations.* These are stations that serve a comparatively small area with a radiated power level up to 3 kW. This power is measured on the basis of effective radiated power (erp); it is equal to the antenna input power multiplied by the antenna gain. If a directional transmitting antenna is utilized, the effective radiated power value denotes the erp value in the direction of maximum gain.

2. *Class B Stations.* These FM broadcast stations serve the larger communities and principal cities. A Class B station may radiate up to 50 kW erp. Antenna heights up to 500 feet may be employed, whereas a Class A station is limited to an antenna height of 300 feet. Both Class A and Class B FM stations may radiate in both the vertical and the horizontal planes. A vertical radiation component enhances sky-wave reception.

3. *Class C Stations.* This is the most powerful class of FM broadcast station. A Class C station may radiate up to 100 kW erp, with an antenna elevation up to 2000 feet. The station may radiate in both the vertical and horizontal planes.

Note in passing that a carrier frequency of 108.0 is assigned by the FCC to VHF omnirange (very-high frequency bearing information), also called VOR test stations. VOR assignments are made only in locations where no interference is caused with FM broadcast transmissions. Note also that 27 of the 100 FM broadcast channels are identified as Class A channels; they are assigned only to Class A stations throughout the country. The first 20 FM channels are reserved for noncommercial educational FM broadcasting.

Following the establishment of basic FM network operations, systems designers turned their attention to *stereophonic sound* transmission and reception. A stereophonic sound system employs two or more microphones, transmission channels, and speakers arranged to provide binaural perception (depth) of the reproduced sound. Stereo FM transmission and reception are characterized by two separate audio signals, left (L) and right (R). This stereo signal is transmitted in a manner that permits stereo reproduction by a stereo FM receiver, or monophonic (mono) reproduction by a monaural FM receiver. A multiplexing technique is utilized to obtain stereo separation in a frequency channel that was originally required for conventional mono operation. *Multiplexing involves a sophistication of basic modulation processes to transmit more than one audio signal on the same RF carrier.* An FM stereo multiplex system employs the modulation spectrum shown in Fig. 2-7. Two microphones (L and R) are energized at the transmitter, and two speakers (L and R) are driven at the receiver.

The L + R portion of the stereo multiplex signal is produced by combining the left and right audio channels that are generated at the studio in phase. In turn, this L + R signal corresponds to a monaural audio signal with a frequency range from 50 Hz to 15 kHz. This portion of the stereo-FM signal can be received and detected as a mono FM transmission by a conventional monaural FM receiver. The L and R signals are also combined, but with the R signal shifted in phase by 180 deg. In other words, the R signal is subtracted from the L signal, and a difference signal (L − R signal) is formed. This L − R signal is used to amplitude-modulate a 38-kHz subcarrier in such manner that two sidebands are produced on either side of the 38-kHz frequency

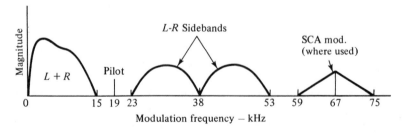

Figure 2-7   Modulation spectrum of the FM stereo-multiplex system.

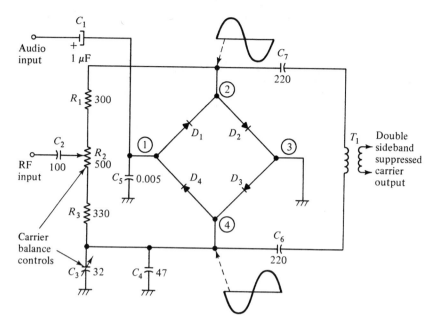

Figure 2-8   A balanced-modulator configuration.

point. However, *the 38-kHz subcarrier frequency is suppressed by means of a balanced modulator* such as that depicted in Fig. 2-8.

The sidebands containing the L − R signal extend from 23 kHz to 53 kHz. This range is supersonic, and is inaudible in this form to the listener. Consequently, a mono receiver effectively reproduces only the L + R signal. As indicated in Fig. 2-7, an unmodulated 19-kHz pilot subcarrier is included in the stereo signal. This pilot signal serves to synchronize the reconstitution of the complete L − R signal in the receiver system. Note that 38 kHz is the second harmonic of the 19-kHz pilot subcarrier. In both mono and stereo FM transmissions, an SCA signal is sometimes included. This is a *Subsidiary Subcarrier Authority* assignment, also called a "storecasting" signal. This SCA modulation extends from 60 kHz to 74 kHz, with a subcarrier frequency of 67 kHz. Unless the SCA signal is trapped out in a home-entertainment receiver, it will interfere with the operation of the stereo decoder circuit and produce an interfering swishing sound.

A stereo FM receiver contains, in addition to an FM detector, a stereo decoder or "multiplex" section that recovers the L and R audio signals from the composite audio signal. This composite stereo signal appears on the screen of an oscilloscope, as shown in Fig. 2-9. Note

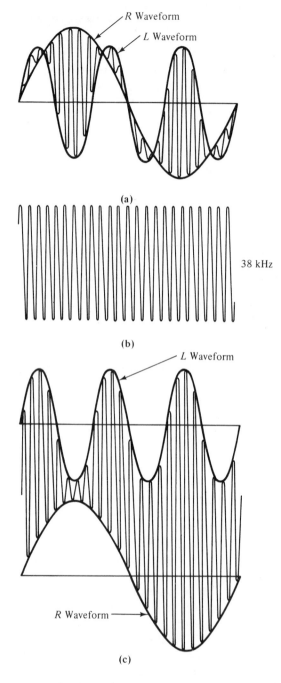

Figure 2-9 Stereo-multiplex subcarrier reinsertion process. **(a)** Incoming L and R signal with suppressed subcarrier; **(b)** locally generated subcarrier; **(c)** L and R signal waveform with subcarrier reinserted.

that the waveform in (a) has a suppressed subcarrier. A locally generated subcarrier (b) is added to this incoming composite signal to form the reconstituted composite signal (c). A significant characteristic of this reconstituted signal is the presence of L information along its positive-peak portion, and the presence of R information along its negative-peak portion. In turn, the L and R signals can be recovered simply by applying the reconstituted signal to a pair of oppositely polarized diode detectors. This is called the envelope-detection method. However, most receivers use another detection method in which reconstitution and detection occur simultaneously in the same circuit.

System designers have also become concerned with *quadraphonic* transmission and reception in FM networks. This mode of information transfer is undergoing development at this writing, and system standards have not yet been established. In quadraphonic broadcasting, the designer is concerned with four separate channels of information: left front, right front, left rear, and right rear. These four signals must be simultaneously transmitted in a channel that originally accommodated only a mono signal. A quadraphonic broadcast system must also be compatible, and thereby reproducible by conventional monophonic and stereophonic receivers. In other words, the quadraphonic transmission must "look like" a monophonic signal to a mono receiver, and must "look like" a stereophonic signal to a stereo receiver. Some system designers are experimenting with arrangements that employ the 67-kHz SCA frequency, either to convey time-multiplexed signals, or to provide four-channel decoding signals to the receiver.

2-4 RADIO BROADCASTING SYSTEM

Most standard broadcast sound sources are tape-system or disc-reproduction arrangements. However, studio facilities are also employed for programming various types of live talent. As exemplified in Fig. 2-10, more than one studio is generally provided, so that one program can be rehearsed while another is being broadcast. Each studio is equipped with several kinds of microphones. The microphones energize pre-amplifiers, which in turn feed into mixer and volume-control facilities installed in a control booth. From the control booth, a line conducts the audio signal to the main control room. After amplification, the audio signal may be fed to a local AM transmitter and/or to distant network transmitters via telephone-company (Telco) lines. A micro-wave studio-transmitter link (STL) unit may be utilized for pickup of external signals, or the main control room may be linked to a transmitter by a microwave unit.

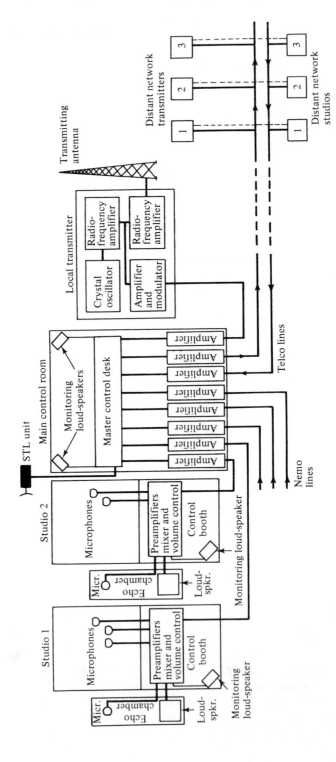

Figure 2-10 Basic arrangement of a radio broadcasting network.

The main control room also has incoming remote (nemo) lines, and incoming lines from other studios. Monitoring loudspeakers are provided in both the control booth and the main control room. A studio is soundproofed, and its walls are designed for control of reverberation. An echo chamber is employed for adding hollow effects or actual echoes to program sounds. *Field events are covered by portable microphones and studio-transmitter links,* often transported in specialized vans. In turn, the audio signals are relayed to the main studio via telephone lines or short-wave radio links. When the event is to be broadcast at some later time, it may be recorded on magnetic tape. In a studio, disc recording is often utilized.

System designers usually support microphones on floor stands, or from overhead cables. Since high-fidelity microphones have a low-level output, preamplifiers are provided in the immediate vicinity; thereby, the audio signal level can be greatly increased before the signal/noise ratio becomes a problem. Each microphone is provided with an individual preamplifier in the control booth adjoining the studio. Some system designers provide for mixing the outputs from a group of microphones, with the mixed output fed to a preamplifier. Note that a typical broadcast microphone provides an average output voltage of 0.5 mV. From the preamplifiers, the audio signals are applied to switching facilities and a *fader* in the control booth. A fader is a multiple-unit control that provides gradual changeover from one microphone to another.

Designers customarily provide a master volume control in the control booth whereby the operator can "ride gain" on the sound level of the program. A volume-indicator (VU) meter and a monitoring loudspeaker are used to check the prevailing signal level. A volume-indicator meter is calibrated in volume units (VU). A *volume unit* is the unit of transmission measurement for nonsteady-state audio voltages. Zero level is defined as the steady-state reference power of 1 mW in 600 ohms. The zero volume unit is equivalent to +4 dBm for a single frequency. Note that dBm is the abbreviation for decibels above (or below) one milliwatt. In other words, it is a power value that is expressed in terms of its ratio to 1 mW.

The central point of the broadcast studio system is termed the main control room. It serves as the distribution center for program sequences from local and remote studios, from studio-transmitter links (STL's) and from remote (nemo) and long-distance (Telco) lines. Audio signals from these various sources are amplified to a standard VU level for line transmission, and are switched according to schedule into lines or short-wave links for transport to the broadcast transmitter. In various situations, the system designer locates the transmitter a number of miles from the studios and control rooms. This decision is governed by considerations of the most desirable location for the studios (such as in a metropolitan area) and the optimum location for the transmitter (as in an unobstructed central site for the region to be served).

Telephone cables and microwave links are generally employed for transporting the audio program signals from a main control room to the broadcast transmitter, or to stations in distant cities in the case of network programming. Telephone companies lease conventional lines and precisely equalized lines to broadcast companies. Most lines are equalized from 100 Hz to 5 kHz. In many systems, carrier transmission is used; carrier frequencies include a very extensive range, from 7.6 kHz up. Nearly all network lines are of the coaxial type; open-wire lines are the exception. Various types of microwave radio relay systems are included in long-distance telephone networks. Transmission is in the super-high-frequency (SHF) range from 3700 to 4200 MHz. Typical transmitting and receiving terminals are depicted in Fig. 2-11.

This 500-MHz band comprises 12 20-MHz channels with 20-MHz separation between channels. Each of these channels is designed to accommodate many audio signals in a carrier system with carrier frequencies in the range from 68 to 2044 kHz. *Repeater stations are installed at approximately 25-mile intervals along radio relay routes.* As indicated in Fig. 2-11, waveguides are employed by system designers for transport of the microwave signals to and from antennas, and between receiver and transmitter devices and subsystems. Antennas utilized in the super-high-frequency range are called *aperture radiators.* This type of antenna consists essentially of a waveguide feed line that delivers electromagnetic wave energy to parabolic reflectors.

The local transmitter, indicated in Fig. 2-10, comprises a crystal oscillator for generation of the RF carrier, RF amplifiers to step up the signal amplitude, an amplifier-modulator arrangement to impress the audio signal upon the RF carrier, and a transmitting antenna for radiation of electromagnetic waves. Efficient radiation is accomplished by utilizing an antenna that has physical dimensions in the same order as the wavelength of the radiated signal. A steel tower is generally employed by system designers, with a vertical height in the range from one-half to one full wavelength. When directional transmission is required, two or more towers are utilized, with a typical separation of one-half wavelength. Vertical polarization is provided, with maximum radiation in the form of a ground wave, and minimum sky-wave radiation.

System design compatibility denotes electromagnetic compatibility achieved by incorporating in all electromagnetic radiating and receiving

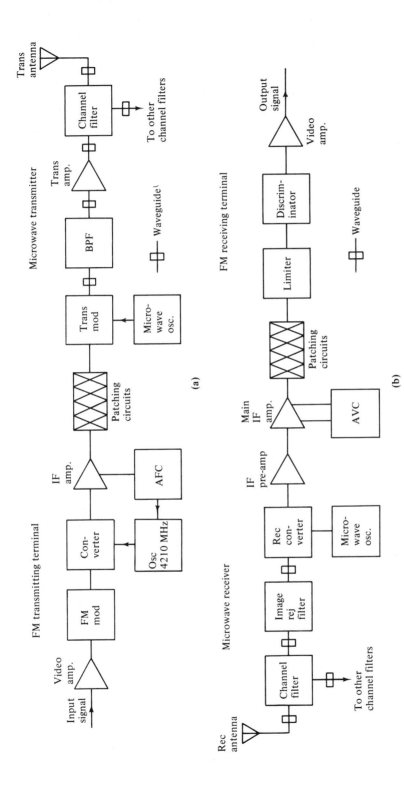

Figure 2-11 Typical microwave radio relay arrangements. **(a)** Transmitting terminal; **(b)** receiving terminal.

65

apparatus, including antennas, characteristics or features for elimination or rejection of undesired self-generated or external signals, for enhancement of operating capabilities in the presence of natural or man-made electromagnetic noise. A *design proof test* is a critical checkout and measurement schedule that is used to verify that a system design specification meets the overall functional requirements of the operating network. This is an aspect of system engineering in which a comprehensive analysis is made of all the elements in the network with respect to signal-processing action.

2-5 CABLE CARRIER SYSTEMS

With reference to Fig. 2-10, the Telco and nemo lines indicated in the diagram are ordinarily coaxial cable installations. Carrier systems are widely utilized. A carrier system is defined as a means of obtaining a number of channels over a single path by modulating each channel upon a different carrier frequency and demodulating at the receiving point to restore the signals to their original form. A carrier line denotes a transmission line, such as a coaxial cable, used for multiple-channel carrier communication. Because the signal voltage becomes progressively attenuated along a cable route, line amplifiers must be provided at intervals. A line amplifier may be designed as a *one-way repeater* or as a *two-way repeater*. A repeater is a combination of apparatus for receiving signals from either one or both directions and delivering corresponding signals that have been amplified and/or equalized. A *carrier repeater* is an assembly, including an amplifier or amplifiers, filters, equalizers, controls, and incidental units, that functions to raise the carrier signal level to a value suitable for traversing a succeeding line section, while maintaining an adequate signal-to-noise ratio.

It is evident that radio transmission systems are similar in various ways to carrier systems. Both systems exploit the frequency-division principle to obtain a multiplicity of separate transmission paths over a single transmission medium. A dozen categories of carrier systems are in use, distinguished principally by the carrier frequencies that are employed and by the bandwidth of the channels. Carrier frequencies range to over 8 MHz. A simplified basic carrier system is depicted in Fig. 2-12. In this example, three carrier frequencies, f_1, f_2, and f_3, are employed, which are amplitude-modulated by the comparatively low signal frequencies. Double-sideband transmission is utilized. At the receiving end of the line, tuned filters are provided to separate the channels prior to demodulation.

If the line is more than several miles in length, repeaters must be

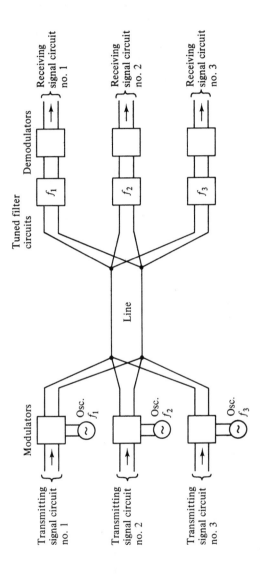

Figure 2-12 A simplified basic carrier system.

included at intervals to maintain an adequate signal-to-noise ratio and to avoid objectionable signal attenuation. For example, a typical telephone cable 1000 miles long imposes a signal loss of almost 500 dB. Thus, an input power of 1 mW at the transmitting terminals would be attenuated to approximately 10^{-51} watt at the receiving terminals. A power amplification of 10^{48} times, approximately, is required to maintain the signal level in this example. Therefore, the system designer must provide repeaters at comparatively frequent intervals along the line. Each repeater may provide a gain of about 15 dB.

A basic two-way repeater arrangement is shown in Fig. 2-13. If a signal flows from the west line into the hybrid coil, it is inductively coupled to amplifier 1, passes through filter 1, into the hybrid coil, and is coupled to the east line. Since the hybrid coil prevents the output from filter 1 from being applied to amplifier 2, the repeater is stable and does not oscillate. Similarly, if a signal flows from the east line into the hybrid coil, it is inductively coupled to amplifier 2, passes through filter 2, into the hybrid coil, and is coupled to the west line. Again, since the hybrid coil prevents the output from filter 2 from being applied to amplifier 1, the repeater is stable and does not oscillate. A hybrid coil is also called a *bridge transformer*. It is a transformer that has, in effect, three windings. The output from a hybrid coil flows into a line and into a balancing network. This balancing network is designed to have the same impedance as the input terminals of the line. In turn, half of the output signal flows into the line and the other half flows

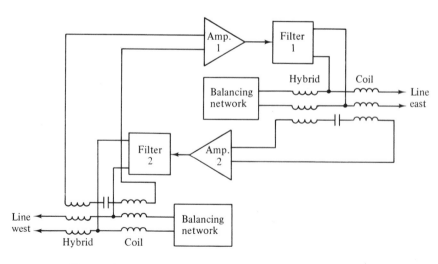

Figure 2-13   Basic two-way repeater arrangement.

into the balancing network. *These outputs cancel in the third winding, so that no input signal is applied to the following amplifier.*

Audio signals in telephone cables have comparatively low power. For example, a sine-wave dialing tone has a source power of approximately 0.3 mW. This power level corresponds to approximately 0.5 V rms in a 900-ohm line. Note that this is a much higher power level than is required to produce an audible sound output from a telephone receiver. Audio signal levels on cables have a source power level that is higher than the level at the load end, owing to line loss. Also, the audio power level applied to a subscriber's telephone set is higher than the power supplied to the receiver because of circuit losses in the subscriber's set. A voice signal with a certain average-power level will have instantaneous power levels that vary from zero to several times the average power level. Peak power levels in speech signals are of very short duration.

Note that the arrangement shown in Fig. 2-13 is called a *two-wire repeater,* because two conductors (or a coaxial cable) are used for signal transmission in either direction. Although two-wire systems are used extensively in radio "talk show" programming, for example, four-wire systems are utilized for high-fidelity programming. A *four-wire repeater* arrangement is conveniently regarded as a development of the configuration shown in Fig. 2-13. In other words, signals traveling in opposite directions are separated. Thus, east-west signals flow through the circuit that includes amplifier 2, whereas west-east signals flow through the circuit that includes amplifier 1. Now, consider that amplifier 1 is to be replaced by a long cable installation that includes a number of one-way amplifiers. Similarly, consider that amplifier 2 is to be replaced by a long cable installation that includes a number of one-way amplifiers. In turn, the hybrid coils and balancing networks are employed *only at the input and output terminals* of these two cable installations.

In effect, a four-wire repeater arrangement has been developed from a two-wire arrangement. One cable installation is then used for transmission in the east-west direction, whereas the other cable installation is used for transmission in the west-east direction. The practical advantage of this design is in the higher gain that is provided by one-way repeaters. In turn, fewer repeaters are required along a route, and smaller conductors can be utilized. A one-way repeater provides considerably more gain than a two-way repeater, because *losses in a balancing network and hybrid-coil subsystem are eliminated.* A four-wire repeater system is also more reliable, because there is no possibility of the subsystem's becoming unbalanced and inoperative owing to self-oscillation. Accordingly, most long-cable telephone networks are designed as four-wire systems.

Elements of
Television System Design

3-1 GENERAL CONSIDERATIONS

A television system includes all the features of a monophonic FM radio broadcast system, plus a complete video network. This video system employs a cable and/or microwave linkage that is generally separate from and often quite differently routed from the audio cable and/or microwave linkage. Thus, the video portion of a network TV program may be transported to the transmitter by a cable from a northeastern terminal, whereas the audio portion of the program may be transported to the transmitter by a cable from a southeastern terminal. Accordingly, *different delay times are often involved in transport of video and audio signals*. However, this is usually a minor consideration in system design, because viewers have an appreciable tolerance for audio timing errors. The most critical situation occurs when a close-up view of a speaker's lips are televised. In this case, the viewer is most likely to sense that the sound signal is preceding or succeeding the video signal. Sometimes, the system designer deems it good practice to include an appropriate delay line in the system.

A map of a partial TV broadcasting system is shown in Fig. 3-1. Line amplifiers (not shown) are provided approximately every four miles. Thus, a 4000-mile video-cable installation includes 1000 line amplifiers. An RG-11/U cable, for example, has an inherent pulse delay of 1.5 μs per 1000 feet. That is, it will require approximately 0.03 sec for a pulse to travel 4000 miles through the cable. If a microwave link were utilized, the transit time would be reduced to approxi-

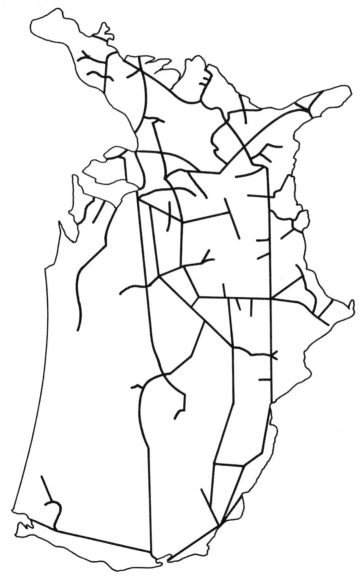

Figure 3-1 Partial TV broadcasting network system. Note: Map shows video cable network; audio cable network is separate.

mately 0.02 sec. A typical studio-transmitter microwave link arrangement is depicted in Fig. 3-2. A microwave link subsystem is also called a *radio relay net*. Microwave amplifier installations (relay transmitters) are employed at 25-mile intervals along a route. Thus, 160 relay transmitters are required along a 4000-mile route.

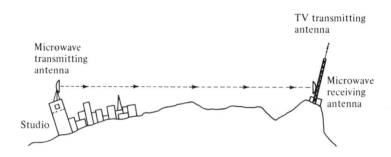

Figure 3-2 Typical studio-transmitter link arrangement.

When microwave transmission is used, equalization for the propagation path is not required. In other words, no frequency distortion or phase distortion occurs as the electromagnetic waves travel through the atmosphere. On the other hand, *when video signals are transported by coaxial cable, equalization is required* to compensate for the high-frequency attenuation and phase shift that takes place in the cable. The system designer also recognizes that the television receivers that are tuned to a transmitted program are also a part of the total system. In other words, *optimum picture reproduction on the receiver screen is obtained only when the signal broadcast from the transmitter has been predistorted in a suitable manner.* Two basic factors are involved in this regard. First, a vestigial-sideband mode of transmission and reception is employed, which impairs the transient response of the receiver subsystem unless compensating predistortion is utilized at the transmitter. Second, a picture tube is nonlinear, and unless suitable gamma predistortion is provided at the transmitter, gray-scale reproduction will be impaired on the picture-tube screen.

System transient response is optimized by provision of suitable phase equalization of the video signal at the transmitter. A predistorted (nonlinear) characteristic is utilized that will cancel the nonlinear phase characteristic in the receiver and provide an overall linear system characteristic, as depicted in Fig. 3-3. If the transient response were not optimized, pulse and square-wave response would be comparatively poor. In other words, a sync pulse, for example, would have excessively slow rise, and its corners would be objectionably rounded. Transitions from white to black or vice versa in the picture image would appear blurred and indistinct on leading edges, and smeared on trailing edges if the system phase characteristic were not linearized by predistortion at the transmitter. This is accomplished in the terminal equipment—the equipment at the end of a communications channel that controls the transmission of signals. Video-frequency cables operate at a typical level of 1 to 2 volts peak-to-peak.

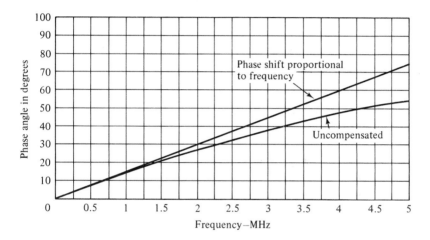

Figure 3-3 Predistortion of the video-signal phase characteristic provides a linear system characteristic.

3-2 SYSTEM DISTORTION PARAMETERS

System designers generally classify system deviations into linear and nonlinear distortion categories. *Linear distortion is defined as a deviation from normal performance that is independent of the signal amplitude.* For example, if a signal channel has subnormal frequency response, and this frequency response remains unchanged from a low signal amplitude to a high signal amplitude, it is termed a linear distortion condition. Again, if a signal channel has an abnormal phase characteristic, and this phase characteristic does not change from low-level operation to high-level operation, it is called a linear distortion condition. On the other hand, if a signal channel has normal frequency response at low signal levels, but develops subnormal frequency response at high signal levels, this malfunction is termed a nonlinear distortion condition. Also, if a signal channel has a normal phase characteristic at high signal levels, but develops an abnormal phase characteristic at low signal levels, this malfunction is called a nonlinear distortion condition.

A simplified representation of a coaxial-cable television network is shown in Fig. 3-4. Since a line amplifier is inserted at each four-mile interval along the cable route, either linear or nonlinear distortion can occur because of line-amplifier deterioration. In addition, temperature changes from winter to summer can cause variations in cable characteristics. System designers employ negative feedback to stabilize line-amplifier operation. Temperature compensation is provided. However, because perfect stabilization and perfect compensation cannot be achieved, there is a continuing possibility that numerous small deviations along a cable route will combine to develop objectionable distortion of the output signal. Both linear and nonlinear distortion can be compensated to a large extent in the terminal equipment. The signal characteristics are monitored by transmitter personnel to determine the corrective processing measures that are desirable.

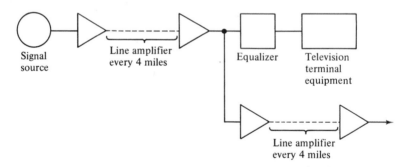

Figure 3-4 Simplified representation of a coaxial-cable television network.

A block diagram of a television transmitter system is exemplified in Fig. 3-5. It is evident that linear or nonlinear distortion can be introduced by marginal operation of the TV camera, the sync distribution amplifiers, the sync generator, the switching amplifier, the film camera, the stabilizing amplifiers, or the transmitter unit. A *stabilizing amplifier* is a signal-processing subsystem that can separate various video-signal components, can shape and reconstruct certain components, and can

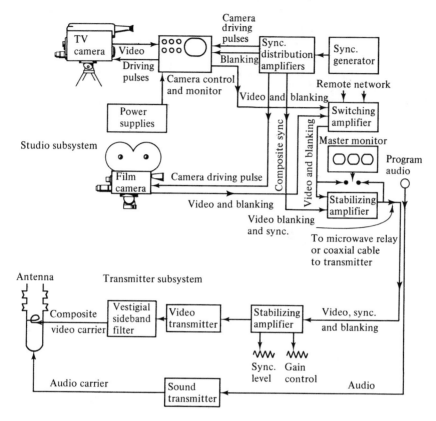

Figure 3-5 Block diagram of a television transmitter system.

reconstitute the video-signal waveform. The camera signal can be processed only to a comparatively limited extent. However, its amplitude will be linearized, if necessary; any spikes extending into the whiter-than-white region will be clipped.

Various forms of interference also enter into system distortion considerations. For example, 60-Hz hum-voltage interference is common. In a square-wave test of a circuit or system that imposes hum interference, the resulting oscilloscope pattern appears as shown in Fig. 3-6. If the square-wave repetition rate is some integral value with respect to 60 Hz, the baseline of the pattern will be sinusoidal. On the other hand, if the square-wave repetition rate is not integrally related to 60 Hz, the horizontal intervals of the display will be thickened. *Square waves are used chiefly to check the transient response of in-*

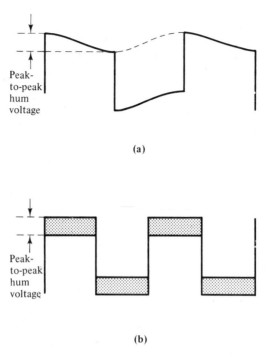

(a)

(b)

Figure 3-6   Basic hum-voltage interference displays. **(a)** Synchronous hum produces sinusoidal waveform envelope; **(b)** nonsynchronous hum produces thickening of horizontal waveform intervals.

dividual amplifiers. However, other types of test signals are preferred for checking the response of the transmitter system. The reason for this preference is that vestigial-sideband transmission is utilized in the high-frequency subsystem. Vestigial sideband reception is also employed in TV receivers. In turn, *system transient distortion* has only an indirect relation to the distortion that is displayed in a simple square-wave test.

Individual amplifiers are periodically checked for correct frequency response by means of a video-frequency sweep-signal test with an oscilloscope to display the frequency response curve. Absorption markers are generally used to identify key frequency points along the response curve. Continuous monitoring of the video-frequency system response up to the point that the monitor oscilloscope is connected is provided by the *standard multiburst signal,* illustrated in Fig. 3-7. This multiburst signal is a portion of the vertical-interval test signal (VITS) that is provided during one horizontal interval of the vertical-retrace period. As noted in the diagram, consecutive multiburst frequencies are 0.5, 1.5, 2.0, 3.0, 3.6, and 4.2 MHz.

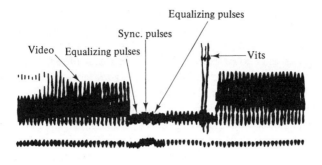

(a)

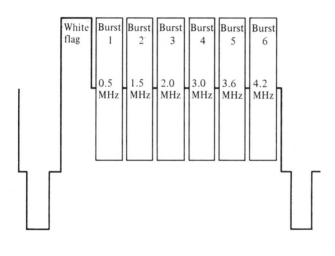

(b)

Figure 3-7 Standard multiburst signal is included in the VITS waveform. (a) Appearance of the VITS waveform in the vertical-retrace interval; (b) expanded view of the multiburst signal, with burst frequencies noted.

A multiburst signal serves the same basic purpose as a sweep-frequency signal. Its chief advantage is that the multiburst signal is introduced into the composite color signal at its point of origin, and accompanies the composite signal to its terminal-equipment point. Thus, *the multiburst signal provides continuous monitoring of the video system's frequency response.* In turn, the transmitting station personnel can make desirable adjustments of signal processing facilities as soon as the requirements occur. Note that the composite video signal in a monochrome transmission consists of the picture signal, blanking signals, and synchronizing signals. In color transmission, color synchronizing signals and color-picture information are added. The composite color signal comprises the color-picture signal plus all of the blanking and synchronizing signals. In other words, it includes the luminance and chrominance signals, the vertical and horizontal sync pulses, the vertical and horizontal blanking pulses, and the color-burst signal. An FM sound signal is associated also with the video transmission.

If the video system has a uniform (flat) frequency response through 4.2 MHz, the six sections of the multiburst signal will be displayed at the same amplitude on an oscilloscope screen. On the other hand, if the higher video frequencies become attenuated in passage through the system, the higher-frequency sections of the multiburst signal will be displayed at comparatively low amplitude, as exemplified in Fig. 3-8. This is an example of linear distortion. After the signal has been equalized, it will be displayed as shown in Fig. 3-9. Of course, *the monitor oscilloscope must have uniform frequency response through 4.2 MHz.* Otherwise, deficiencies in oscilloscope response would be attributed to the system under test.

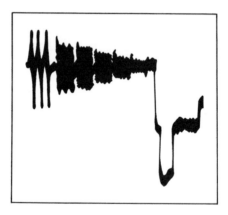

Figure 3-8   Example of high-frequency attenuation in a multiburst signal.

It is evident that a multiburst signal cannot provide an indication of system amplitude nonlinearity. Neither can a square-wave signal provide useful information concerning system amplitude nonlinearity. Therefore, the system designer includes a vertical-interval test signal that indicates any departure from amplitude linearity in system operation up to the point of connection of the monitor oscilloscope. The most

Figure 3-9 Display of multiburst signal with full bandwidth in the system under test.

widely used test signal is the staircase waveform depicted in Fig. 3-10. This signal has 11 discrete levels; in most cases, a 3.58-MHz burst is superimposed on each step. Note that a simple ramp (sawtooth waveform) is sometimes used instead of a staircase waveform for checking amplitude linearity. If the system is linear, the waveform is reproduced as shown in Fig. 3-10. On the other hand, if amplitude nonlinearity occurs, the ramp will become curved up or down.

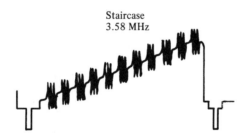

Figure 3-10 Staircase / 3.58-MHz signal used for checking system amplitude linearity.

Note that the 3.58-MHz bursts superimposed on successive levels of the staircase waveform provide a critical means for checking system linearity. In other words, if any amplitude nonlinearity is present, the bursts will have unequal peak-to-peak voltages. At the transmitting terminal, the staircase/3.58-MHz signal is often processed through a high-pass filter, and the filter output is fed to the monitor oscilloscope. In turn, 11 successive bursts are displayed along a horizontal base line (the ramp component has been filtered out). In turn, any small variations in peak-to-peak voltages of successive bursts become clearly apparent to the operator, who can promptly take remedial steps to compensate for the amplitude nonlinearity.

To check for departures in system transient response from optimum, system designers generally provide sine-squared (sin²) pulse test facilities. This is a special waveform that is included in the VITS group of waveforms. A sine-squared waveform is formed when a sine wave is multiplied by itself, as depicted in Fig. 3-11. The product has twice the frequency of its factors. Note also that a sine wave has equal positive and negative excursions, whereas a sine-squared wave has a positive excursion only. In other words, a sine-squared waveform consists of a sine wave with a DC component equal to the peak value of the sine wave. Specifications for a sin² pulse are shown in Fig. 3-12. The pulse width is called its *half-amplitude duration* (h.a.d.); it is generally stated in T units, where T is the duration or period of one picture element in the system. In a 4-MHz video system, T is equal to 0.125 μs.

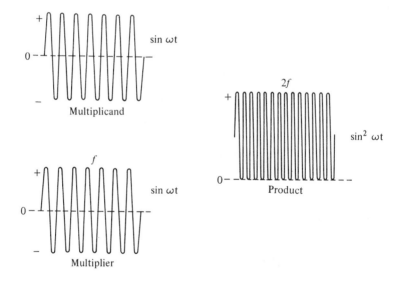

Figure 3-11   A sine-squared waveform is produced by multiplication of a sine wave by itself.

An important advantage of a sin² T pulse, or T pulse, is that *its waveform is practically the same as that of a video signal corresponding to a picture element*. Therefore, when a T pulse passes through a system with minimum distortion, a high-frequency video signal will also pass through the system with minimum distortion. The appearance of a sin²

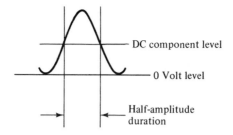

Figure 3-12   Specifications of a sin² pulse.

pulse in a horizontal line of VITS information is shown in Fig. 3-13. Although the pulse appears highly compressed with respect to a complete horizontal line, when it is expanded on the screen of an oscilloscope with high-speed time-base facilities, the pulse will then be seen to have the sinusoidal shape depicted in Fig. 3-12. Note that the sin² pulse is followed by a window pulse in Fig. 3-13. The window pulse provides the operator with a peak-to-peak voltage reference for low-frequency signals, to indicate whether the high-frequency pulse may have been attenuated in passage through the system.

A system may be tested for transient response with a $2T$ pulse. In a 4-MHz system, a $2T$ pulse will have a half-amplitude duration of 0.25 μs. A $2T$ pulse has fewer high-frequency harmonics than a T pulse and is used to check transient response at lower frequencies. System designers also specify the use of a $20T$ pulse; in a 4-MHz system, a $20T$ pulse has a half-amplitude duration of 2.5 μs. This $20T$ pulse is used to amplitude-modulate the 3.58-MHz color subcarrier in a color-TV system. *Its purpose is to check the difference in gain that may be present between the low-frequency end and the high-frequency end of the video-frequency spectrum.* It is also used to check the phase relation (delay time) between signals at the high and low ends of the video-frequency spectrum. These checks are made by observing the amount and kind of distortion displayed in the reproduced pulse after passage through the system.

In addition to providing the operator with a peak reference for low-frequency signals, the window pulse will indicate low-frequency linear distortion in terms of undershoot, tilt, or overshoot. Undershoot and overshoot are related to phase irregularities, and tilt is related to impaired low-frequency response. As noted previously, the transient response in a vestigial-sideband system differs from that in a double-sideband system, so that predistortion of the video signal is required at the transmitter in order to achieve optimum system response. This

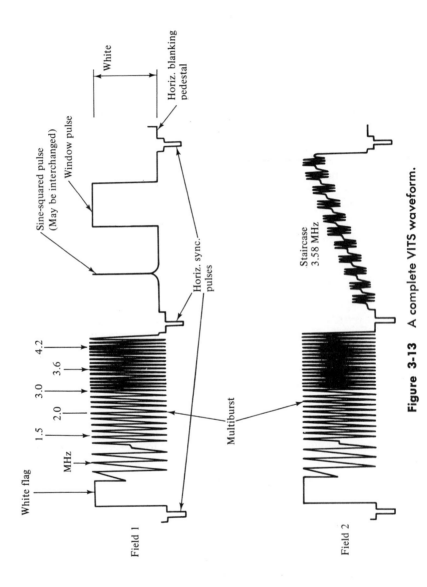

Figure 3-13 A complete VITS waveform.

respoise is shown in Fig. 3-14. When a step response (such as a window pulse) is applied to a 4-MHz VSB system, the reproduced step response at the receiver will be distorted; it will exhibit undershoot, slow rise, smear (tilt), and ringing. After phase equalization and amplitude compensation are introduced at the transmitter, the reproduced step response at the receiver becomes optimized, although it is not distortionless. This optimized response has equal ringing intervals at its black and white levels, a comparatively fast rise, and no smear interval (tilt).

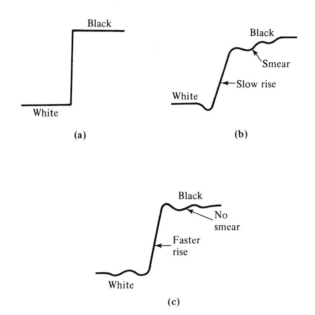

Figure 3-14   Step response in a VSB 4-MHz system. **(a)** Ideal step voltage; **(b)** system response without predistortion; **(c)** system response with predistortion.

Next, consider video-system checks with T pulses. With reference to Fig. 3-15, if a T pulse is passed through a system that has incorrect frequency response, but correct phase response, a typical form of distortion is displayed by the reproduced pulse. *This distortion consists of lobes preceding and following the main excursion of the reproduced T pulse.* These lobes consist of undershoots and ringing excursions. Because the leading lobes and the trailing lobes are the same in Fig. 3-15(b), it is indicated that the system phase response is correct and that only the frequency response is incorrect. Note that if the system phase response and frequency response are both correct, a reproduced

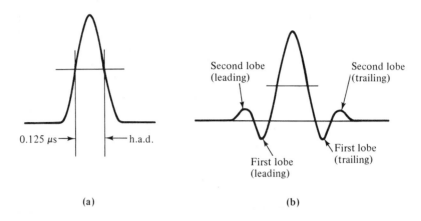

(a)   (b)

Figure 3-15   Basic *T*-pulse (sin² pulse) response, 4-MHz system. (**a**) Input *T* pulse; (**b**) output *T* pulse showing incorrect frequency response and correct phase response.

T pulse will still show a first-lobe development. The reason for this is that the frequency spectrum of a *T* pulse extends past the 4-MHz cutoff point. Thus, it is the second-lobe development in Fig. 3-15(b) that indicates incorrect frequency response of the system.

System designers also provide 2*T* pulses for operational checks. A 2*T* pulse in a 4-MHz system has a half-amplitude duration of 0.25 μs. The basic usefulness of the 2*T* pulse is that its frequency spectrum does not extend past the 4-MHz cutoff point of the transmission channel. In turn, if the system has normal frequency response, the 2*T* pulse is reproduced in undistorted form without any first-lobe development. However, a 2*T* pulse test has a limitation, in that it does not clearly indicate any abnormalities toward the high-frequency end of the transmission channel. It will indicate abnormalities up to approximately 75 percent of the channel cutoff frequency, and provides no useful information concerning the final 25 percent of high-frequency channel response. Therefore, it is good practice to check video-channel response both with a *T* pulse and with a 2*T* pulse.

Next, consider the indication of system phase abnormalities by a *T* pulse. When the system phase characteristic is nonlinear, the reproduced *T* pulse is not symmetrical. Instead, the pattern becomes skewed, either toward the leading edge or toward the lagging edge. In addition, as depicted in Fig. 3-16, an unsymmetrical lobe development occurs. In other words, lobes at the left end of the pattern indicate high-frequency lead in the system; on the other hand, lobes at the right end of

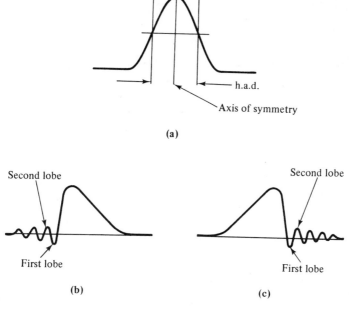

Figure 3-16 Nonlinear phase indication by a *T* pulse. **(a)** Normal *T* pulse display; **(b)** system high frequencies leading; **(c)** system high frequencies lagging.

the pattern indicate high-frequency lag in the system. These distortions are corrected by means of phase equalizers. Phase equalization does not eliminate lobe development. However, it serves to make the reproduced *T* pulse symmetrical, to reduce the lobe amplitude, and to distribute the lobe development equally at the left and right ends of the pattern.

High-frequency attenuation in the system causes a T pulse to reproduce at subnormal amplitude, and also increases the width of the reproduced pulse. If the high-frequency attenuation is gradual, these effects on the *T* pulse are not accompanied by ringing (lobe development). On the other hand, a rapid attenuation of the high frequencies causes considerable ringing, with only a minor reduction in pulse amplitude. The ringing frequency is determined by the frequency at which the frequency-response curve drops substantially. For example, in a conventional frequency-response curve with rapid attenuation at its high-frequency end, the ringing frequency is determined by the cutoff frequency (point at which the response is 3 dB down). It is not a common occurrence for the video system to develop a high-frequency rise, or hump. However, in the event that this abnormality should occur, the ringing frequency will be determined by the hump frequency in the response curve.

3-3   NONLINEAR DISTORTION CHARACTERISTICS

As noted previously, transcontinental networks are in general use and are in a progressive state of development. A telephone cable that includes eight coaxial cables is depicted in Fig. 3-17. Each coaxial cable can transport one television program in one direction (cable operation is not reversible). As indicated previously, carrier transmission is utilized, with frequencies extending up to 8.5 MHz. Since installations up to 4000 miles of cable are employed, there is a transmission loss of approximately 40,000 dB that must be offset by the provision of approximately 1000 line amplifiers. Because line losses are greater at higher frequencies, equalizers must be included to maintain the frequency response at a uniform level within 0.25 dB, approximately. Inasmuch as nonuniform frequency response is associated with nonlinear phase response, equalizers must also be included to maintain the differential envelope delay within 0.1 μs, approximately. In other words, a high-frequency video component will not lag or lead a low-frequency video component by more than 0.1 μs.

Figure 3-17   Example of a telephone cable that includes eight coaxial cables.

Because of residual system nonlinearities, system designers must contend with distortion characteristics of differential gain and of differential phase. Differential gain is defined as the ratio of the output amplitudes of a small high-frequency sine-wave signal at two stipulated DC component levels, this ratio being subtracted from unity. In other words, this difference is the value of the differential gain under the stated test conditions. Differential phase is defined as the difference in phase shift through a television system for a small, high-frequency sine-wave signal at two stipulated DC component levels. It follows that if the transfer characteristic of a system is nonlinear, both differential-gain and differential-phase distortion will occur. Differential gain is generally measured with a staircase signal, as provided in the VITS waveform. Differential phase is generally measured with a vectorscope or equivalent phase-indicating instrument.

Differential gain is undesirable in a video signal because the gray range becomes distorted. If the transfer characteristic is linear, as depicted in Fig. 3-18(a), the gray range is correct. On the other hand, if white compression is present, the lighter grays are reproduced as white. Again, if black stretch occurs, the darker grays are reproduced as black. If both white stretch and black stretch are included in the transfer characteristic, only medium grays are reproduced correctly; lighter grays are reproduced as white, and darker grays are reproduced as black. If white compression is present, the lighter grays are reproduced as white. If black compression occurs, the darker grays are reproduced as black. When both white compression and black compression are included in the transfer characteristic, darker grays are reproduced as black and lighter grays are reproduced as white; only medium grays are reproduced correctly. The distorting effect of black compression on a staircase signal is seen in Fig. 3-19.

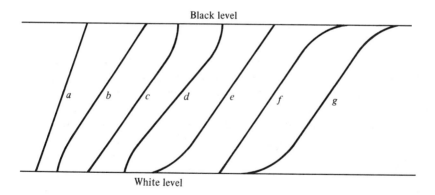

Figure 3-18 Basic transfer characteristics. (a) Linear; (b) white stretch; (c) black stretch; (d) white and black stretch; (e) white compression; (f) black compression; (g) white and black compression.

Nonlinear transfer characteristics are compensated at the terminal equipment by passing the video signal through a nonlinear amplifier that is adjusted to have an opposing curvature in its transfer characteristic.

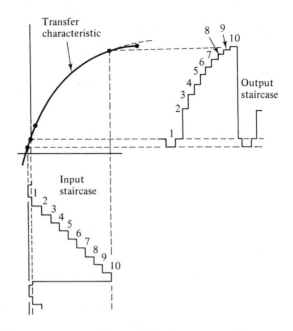

Figure 3-19 Distorting effect of black compression on a staircase signal.

Correct compensation is indicated by the transmission of an undistorted staircase signal. After the transfer characteristic has been linearized, the resulting video signal is checked for linearity of its phase characteristic. Differential phase distortion is particularly undesirable in a color-TV signal, because it makes hue dependent upon the brightness level. For example, if an actress who is wearing a yellow dress walks across the stage from one illumination level to another illumination level, the change in background brightness may cause the apparent color of her dress to change from yellow to green-yellow, and then to orange. To avoid this type of hue distortion, the differential phase error of the system is minimized by adjustment of a compensating nonlinear phase equalizer.

4

Radar System
Design Fundamentals

4-1 GENERAL CONSIDERATIONS

Radar systems form a specialized branch of electromagnetic wave
technology based on the principles of radio transmission and reception.
A radar installation may be a subsystem within an operations system, as
exemplified in Fig. 4-1. *The basic function of radar is to detect the
presence of objects, to determine their direction and range, and to
recognize their character.* This detection is accomplished by radiating a
beam of radio-frequency energy over a region to be searched. When
the beam strikes a reflecting object, electromagnetic wave energy is
reradiated. A very small portion of this reradiated energy is returned to
the radar system. A sensitive receiver located near the transmitter can
detect the echo signal and thereby indicate the presence of an object
or target. Determination of the actual range and direction is based on
the facts that radio-frequency energy travels at the constant velocity of
light and that the receiving system can be designed as a directional
indicator.

   Electromagnetic waves may be reflected from the ionosphere, or
they may penetrate the ionosphere, as depicted in Fig. 4-2. Whether a
beam is reflected from, or penetrates through, the ionosphere, depends
upon its frequency and also upon its angle of radiation. In other words,
if the beam has a frequency that permits it to be reflected from the
ionosphere when radiated at a suitable angle with respect to the vertical
direction, it will nevertheless penetrate the ionosphere when radiated
at an elevation greater than the critical angle. When echoes are to be

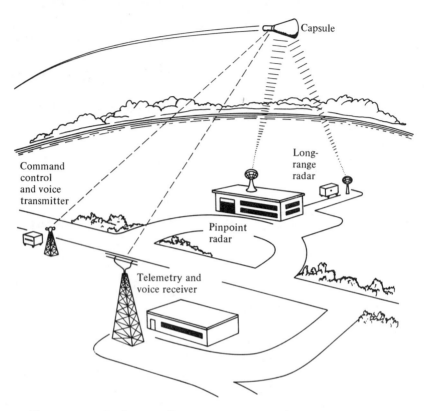

Figure 4-1　Radar installation in a spacecraft operations system.

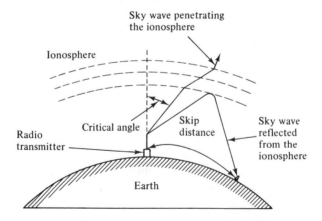

Figure 4-2　Example of electromagnetic wave reflection and of penetration.

received from the moon, or from a capsule in outer space, it is necessary, of course, that the radar beam penetrate the ionosphere. Conventional radar operations are allocated various bands, with frequencies
from 300 MHz to 40,000 MHz, as listed in Table 4-1. These frequencies
are comparatively high, and are not reflected by the ionosphere to any
practical extent.

TABLE 4-1

Conventional Radar Frequency Bands

Band	Center Frequency	Wavelength	
P	300 MHz	1	meter
L	900 MHz	33	cm
S	3,000 MHz	10	cm
C	5,000 MHz	6	cm
X	10,000 MHz	3	cm
K	20,000 MHz	1.5	cm
Q	40,000 MHz	0.75	cm

*Radar systems usually operate with a pulse-modulated carrier;
pulse widths range from 1 to 50 μs.* If the transmitter completes the
radiation of a pulse before the reflected energy returns from a reflecting
object, the receiver can distinguish between the transmitted pulse and
the reflected pulse. After the receiver completes the processing of the
incoming pulse, the transmitter may radiate a second pulse of electromagnetic wave energy, and the process is repeated. Output from the
receiver is applied to an indicator that measures the time interval between the transmission of a pulse and the return of a reflected pulse.
In turn, this time interval is a measure of the distance, or range, over
which the pulse has traveled. When a pulse of electromagnetic wave
energy strikes a reflecting object, its energy is redirected instantly, without any loss of time. Wave travel occurs at the speed of light: 186,000
miles per second, or 328 yards per microsecond.

A cathode-ray tube is generally used as a display indicator. If a
linear sweep (horizontal beam deflection) is used, each horizontal interval represents a certain value of elapsed time, and in turn, the range
indicated by the reflected pulse. If there are several reflections from
various objects, the returned pulses will be displayed at various horizontal intervals that correspond to the ranges of the pulses. Another dimen-
sion is required to locate an object in space. This is its angle of

elevation, or altitude. If either its angle of elevation or its altitude is known, the other can be calculated from the right-triangle relationship and the slant range, as depicted in Fig. 4-3. As pictured in Fig. 4-1, directional antennas are employed, so that the angle of elevation is precisely indicated.

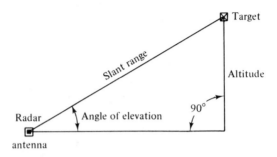

Figure 4-3   Determination of altitude.

4-2   SYSTEM ORGANIZATION

A functional breakdown of a pulse-modulated radar system resolves itself into six fundamental components. With reference to Fig. 4-4, these components are described as follows:

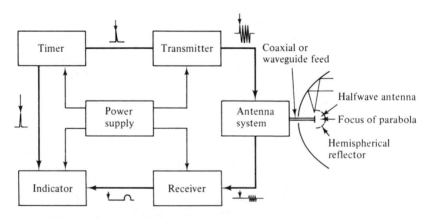

Figure 4-4   Functional black diagram of a basic radar system.

1. The *timer,* also termed a synchronizer, keyer, or control central, supplies the synchronizing pulses that time the transmitted pulses and the indicator, and which coordinate other associated circuits.

2. The *transmitter* generates RF energy in the form of high-energy pulses.

3. The *antenna system* transports the RF energy from the transmitter, radiates it in a highly directional beam, receives any returning echoes, and feeds these echoes to the receiver.

4. The *receiver* amplifies the weak RF pulses returned from the reflecting object and reproduces them as video-frequency pulses that are applied in turn to the indicator.

5. The *indicator* produces a visual display of the echo pulses in a manner that provides the necessary range and identification information.

6. The *power supply* furnishes all AC and DC voltages that are necessary for operation of the system components.

Any radar system has certain associated constants. *A designer's choice of constants for a particular system depends upon its application, required accuracy, range to be covered, physical size considerations, and problems of generating and receiving the electromagnetic wave signal.* Consider, for example, the carrier frequency stipulation. The carrier frequency is the transmitter and receiver operating frequency. Principal factors influencing a selection of carrier frequency are the desired directivity, means of RF energy generation, and means of reflected wave reception. For optimum determination of direction of the reflected energy, and for concentration of the transmitted energy from the viewpoint of efficiency, a radar antenna should be designed for high directivity.

A high carrier frequency corresponds to a short wavelength; if a shorter wavelength is utilized, a smaller antenna array can be used to obtain a specified sharpness of pattern (a radiating element is ordinarily a half-wave in length). For an antenna of a given physical size, a sharper radiation pattern will be obtained at a higher operating frequency. The problem of generating and amplifying reasonable amounts of electrical energy at extremely high frequencies is complicated by the reduced efficiency of electron devices at extremely high frequency (EHF). The lowest carrier frequency is generally considered to be 100 MHz, in view of limiting the antenna array to a practical size and yet to obtain a sufficiently directional beam. *Some designers employ an operating frequency of 10,000 MHz, and higher, in order to obtain very narrow beams and/or to reduce antenna size.*

Sufficient time must be allowed between transmitted pulses for an echo to return from any object within the maximum workable range of the system. Unless this precaution is observed, reception of echoes from the more distant objects will be obscured by succeeding transmitted pulses. This necessary time interval establishes the highest frequency that can be used for pulse repetition. When the antenna is rotated at a constant speed, the beam strikes an object for a comparatively short time. During this time, *a sufficient number of pulses must be transmitted that the returned signal will produce a useful indication on the indicator screen.* Thus, persistence of the screen phosphor and the rotational speed of the antenna determine the lowest pulse repetition rate that can be utilized. In a system wherein the entire interval between transmitted pulses is processed by the indicator, the repetition rate must be very stable to provide accurate range measurement. The minimum range at which an object can be observed is determined chiefly by the pulse width; if an echo is returned before the transmitted pulse ends, echo reception will be masked by the transmitted pulse.

4-3   TIMER AND TRANSMITTER

A timer has the function of ensuring that all circuits associated with the radar system operate in a definite time relationship with one another and that the interval between pulses is of suitable length. Designers usually choose one of two established methods of timer operation. The pulse repetition rate can be determined by any stable oscillator, such as a sine-wave oscillator, a multivibrator, or a blocking oscillator. This oscillator output is then applied to appropriate pulse-shaping circuits to form the required timing-pulse waveshape. Typical radar timing arrangements are depicted in Fig. 4-5. The timing of associated circuits can be accomplished by the output from the timer or by obtaining a timing signal from the transmitter as it is turned on. A transmitter, with its associated circuits, may establish its own pulse width and pulse-repetition rate, and provide a synchronizing pulse for other components in the system. This action may be accomplished by a self-pulsing or blocking RF oscillator with properly chosen circuit constants. This method of timing eliminates a number of special timing circuits, but the pulse width or pulse-repetition rate may in turn be less rigidly controlled than is desirable in various applications.

A self-pulsing radar transmitter provides the functions of transmitting and timing by one component, as seen in Fig. 4-6. This type of transmitter oscillates, in effect, at two frequencies: the *carrier frequency,* as determined by the LC constants of the tank circuit, and the *pulsing frequency,* as determined by the RC time-constant in the device (or tube) control circuit. However, in the externally pulsed type of radar transmitter (Fig. 4-7), the function of the RF oscillator is compara-

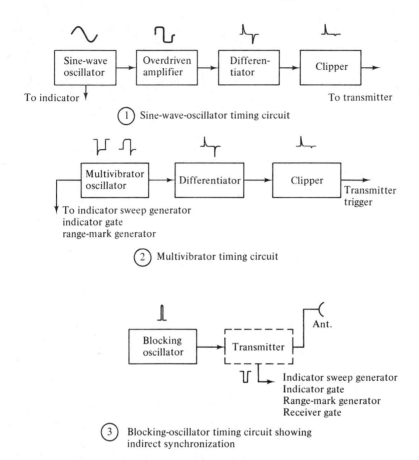

Figure 4-5 Basic radar timing arrangements.

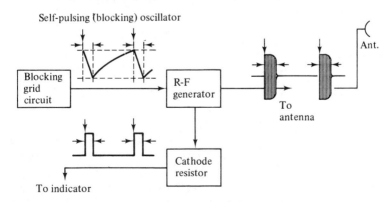

Figure 4-6 A self-pulsing oscillator functions as a transmitter and as a timer.

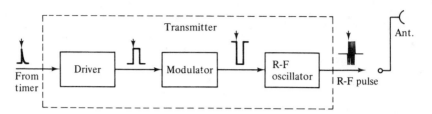

Figure 4-7   Plan of an externally pulsed radar transmitter.

tively simple; it merely generates powerful pulses of RF energy at regular intervals. Since the resting time is lengthy compared to the transmitting time, the oscillator may be greatly overloaded during transmission to increase the peak-power value. In this type of operation, the RF oscillator requires power in the form of a properly timed, high-amplitude rectangular pulse. In most cases, the timing oscillator cannot meet this requirement directly, and it accordingly becomes necessary to include a driver and a modulator.

A driver is a configuration that, when triggered, drives the modulator with a rectangular pulse that has an accurately timed width. The modulator supplies power to the RF oscillator in the form of a timed, high-amplitude, rectangular pulse. The driver is triggered by the timer in order to maintain correct repetition rate of the system. When triggered, the driver shapes a pulse with proper time duration, which in turn operates the modulator. In response, the modulator furnishes high plate voltage to the RF oscillator for a predetermined pulsing time. Accordingly, the transmitting function may be accomplished by the combined action of a driver, a modulator, and an RF oscillator. The modulator operates as a power amplifier from the driver and as a switch for the RF oscillator.

4-4   ANTENNA

A radar antenna functions to transport high-frequency electrical energy from the transmitter to a suitable site, to radiate this energy in a directional beam, to pick up the returning echo, and to conduct it to the receiver with minimum loss. Thus, *the antenna is considered to include the transmission lines from the transmitter to the antenna array, the antenna itself, the transmission line from the antenna array to the receiver, and any antenna-switching device and receiver-protective device that may be included by the designer.* When a radar receiver is operated in close proximity to a powerful radar transmitter, a certain amount of the transmitter signal inevitably gains entry into the receiver by way of the stray capacitance of the input circuit leads. In some instances, such signals from the transmitted pulse must be entirely eliminated from the output of the receiver. Therefore, the receiver must be gated, or turned off, during the pulse time, so that it is made completely unresponsive.

Sometimes it is desirable to couple a small amount of the transmitted RF energy into the receiver for timing purposes. However, the signal directly available from the transmission line is so strong that the receiver is likely to be burned out, or the circuits may be temporarily blocked. These considerations place a limit on the permissible amplitude of pulse that can be coupled from the transmitter into the receiver. A receiver protective device is designed to provide this function. An elementary radar antenna system would employ two separate antennas: one for transmitting, and one for receiving, as depicted in Fig. 4-8. In this arrangement, the receiving antenna must be shielded from the transmitting antenna to protect the receiver from the powerful pulses of radiated energy. Usually, the directivity of the antennas is sufficiently great to permit the location of the receiving antenna in a region of minimum signal strength from the transmitting antenna.

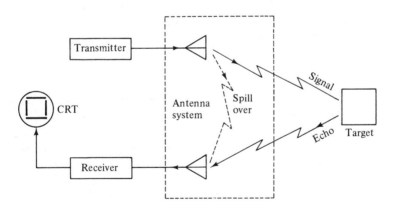

Figure 4-8 Basic antenna system with separate antennas for transmitting and receiving.

A more practical radar antenna design employs a single antenna and an antenna switch that connects the antenna to the transmitter during transmission time, and connects the antenna to the receiver during the remainder of the pulse cycle (see Fig. 4-9). This transmit-

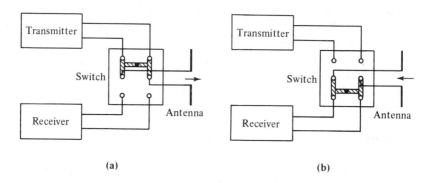

(a)   (b)

Figure 4-9   Single antenna arrangement with transmit-receive (TR) switch. **(a)** Antenna transmitting; **(b)** antenna receiving.

receive (TR) switch is necessary to protect the receiver from the high energy output of the transmitter during the pulse time, and also to isolate the transmitter during the receiving time. Otherwise, the weak echoes might be partially or wholly lost in traversing a transmission line back to the transmitter. The transmitted pulse and the repetition rate of the system, which may range from 60 to 4000 pulses per second, is too rapid to be mechanically switched. Therefore, a resonant-cavity electronic switch is utilized, as depicted in Fig. 4-10. This arrangement is detailed subsequently.

Efficiency is a primary consideration when one is designing a single antenna for both transmission and reception. In other words, all of the energy supplied by the transmitter should be transported to the antenna, and all of the return-echo energy intercepted by the antenna should be transported to the receiver. Optimum efficiency is obtained by matching the impedance of the antenna to the impedance of the transmission line. During transmission of a pulse, the transmitter should be matched to the transmission line, whereas the receiver must present an open circuit or high impedance to the transmission line. Next, during the reception time, the impedance relations should be reversed. The problem of switching is usually simplified because most transmitters have a different impedance when they are on, compared with their impedance when they are off. If the transmitter is properly matched to the transmission line during pulse time, the transmitter will be mismatched during receiving time, and the transmission line will become resonant. A typical elementary system in which the transmitter and receiver are connected to branch lines and thence to the antenna feed line is shown in Fig. 4-11.

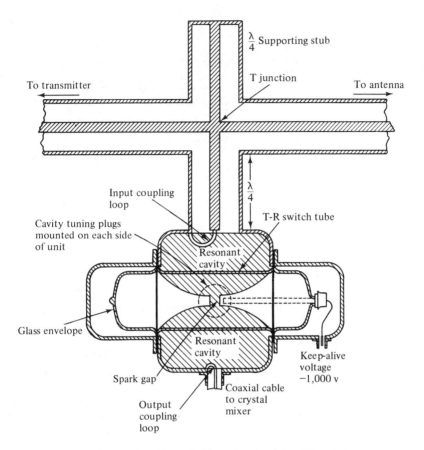

Figure 4-10   Plan of a resonant-cavity TR switch.

The junction of the three lines depicted in Fig. 4-11 is termed the T junction. During the off period, the switch in the receiver branch is closed, and the transmission line from the antenna to the receiver is properly matched. The resistance seen from the T junction looking toward the transmitter can be controlled by the length of the resonant section between them. If the transmitter impedance decreases when it is turned off, the length should be a quarter-wavelength, or some odd multiple thereof, in order to present a high impedance. The high impedance presented by the transmitter and its feed line to the T junction is in parallel with the relatively low characteristic impedance of the remainder of the transmission line system, but, being high, has little effect. If the transmitter impedance increases when it is turned off, the

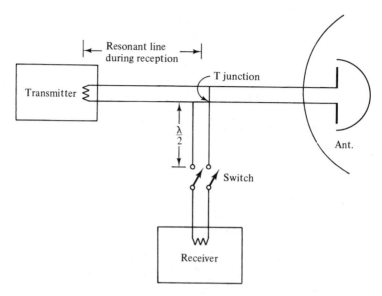

Figure 4-11   Basic TR switching arrangement.

resonant-line length should be a half-wavelength, or a multiple thereof.

When the transmitter is turned on to transmit the next pulse, it will again be properly matched to the antenna. The open switch in Fig. 4-11 will prevent the pulse from reaching the receiver, and will cause a mismatch to the line between the switch and T junction. By utilizing some multiple of a half-wavelength, the open circuit of the switch will be presented as an open circuit across the transmitter-antenna line. The basic design problem in analysis of a switch arrangement consists of providing what amounts to a double-pole single-throw switch for connecting the antenna alternately to the transmitter and to the receiver. *This switching device must be capable of acting within a time interval of a few microseconds,* as the receiver should be connected into the antenna circuit immediately after the transmission of the pulse in order to detect close-range objects.

A parabolic-reflector type of antenna is pictured in Fig. 4-12. It provides a practical means of producing a narrow beam pattern in the microwavelength region. The reflection of RF energy by the parabola, or *dish,* is analogous to the reflection of light by a parabolic mirror. This dish is large in comparison with the operating wavelength; in general, the larger the reflector, the narrower the beam pattern. RF energy is fed to a dipole located at the focal point of the parabola. A parasitic reflector is placed approximately one-quarter wavelength in

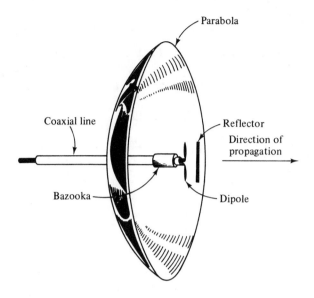

Figure 4-12 Parabolic reflector with dipole antenna design.

front of the dipole, to reflect practically all of the radiated energy back to the dish from which it is reflected ahead in a narrow beam.

4-5 RECEIVER FUNDAMENTALS

A block diagram for a basic radar receiver is shown in Fig. 4-13. Super-heterodyne circuitry is utilized, with seven IF amplifier stages. Improved receiver design can increase the usefulness of radar equipment, perhaps more than any other single factor. Only a small part of the energy radiated from the antenna strikes a distant object, despite the use of a narrow beam of transmitted electromagnetic wave energy. Reflections from the object are scattered in random directions, causing the echo that returns to the receiver to be very weak. In turn, the receiver must accept signals that are often a microvolt or less in amplitude, and amplify them to a suitable level for display on the screen of a cathode-ray tube. Thus, the effective range of a radar installation is proportional to the ability of its receiver to process weak incoming signals.

Signal-to-noise ratio is the primary design parameter. Theoretically, it is possible by use of many stages of amplification to raise any signal, no matter how weak, to any desired amplitude. However, vari-

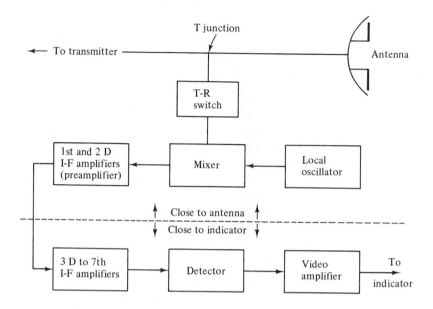

Figure 4-13   Block diagram for a basic radar receiver.

ous random disturbances are present in all electrical circuits that produce small voltage variations termed noise. In particular, the input stage of a receiver generates noise voltages that are most apparent in the output from the receiver. In other words, these noise voltages are amplified along with the signal voltage in the following stages of the receiver. Noises are generated in other stages, but these are less important, since they are not amplified as much as those that occur in the first stage. If the signal is not at least as large in amplitude as the noise voltage at the input stage, this signal cannot be recognized at the receiver output and is therefore useless.

A primary consideration in receiver design is to keep the noise level as low as possible so that, for a given signal, the signal-to-noise ratio is high. If the noise level is low, a weak signal from a distant object may be detected. On the other hand, if the noise level is high, the reflecting object must be much closer before its echo is sufficiently strong to override the noise level. Thus, the noise generated in the receiver is a factor that affects the useful range of the receiver. Noise reduction by improved design in a moderately good receiver may extend the range of the system much more effectively than by increasing the power output from the transmitter. Noise voltages generated in an amplifier stage include three types: thermal agitation, shot or partition noise, and induced noise. All of these comprise frequency components throughout the entire frequency spectrum, and, in turn, the noise amplitude is a function of receiver bandwidth.

Refer to Fig. 4-14. In this example, a UHF superheterodyne receiver of conventional wide-band design for pulse reception is depicted. The receiver components consist of a klystron local oscillator, a crystal mixer mounted in a resonant cavity, six stages of IF amplification, a diode detector, a video amplifier, a cathode follower, and the power supplies. A schematic diagram is shown in Fig. 4-15. The chief operating features of this arrangement are high gain combined with high signal-to-noise ratio, short recovery time, ease of adjustment, and employment of a separate mixer-amplifier section ahead of the main unit. This latter feature permits the mixer stage to be located near the antenna with the receiver output stage located close to the indicator.

At 3000 MHz (10 cm) the echo pulse is applied directly to the mixer, without prior RF amplification. The mixer must be located as near the antenna as possible to minimize transmission-line losses. The position of the antenna is such that, as a rule, *it is impractical to mount the entire receiver near the antenna.* The mixer and local oscillator are placed in the immediate vicinity of the antenna in this arrangement so that the weak echo signals can be converted to the 30-MHz IF frequency before they are attenuated appreciably by the transmission line. The first two IF stages follow immediately as a preamplifier to prevent the IF signal from being lost owing to attenuation in the coaxial feed line to the remotely located receiver chassis. Several factors must be taken into consideration. The incoming echo signal has a very small amplitude, usually in the order of a few microvolts. In turn, the receiver must have a very high gain. Because of the low amplitude of the echo pulse, a high signal-to-noise ratio must be maintained so that weak echoes are not obscured by internally generated noise.

In selecting the receiver bandwidth, the designer considers the waveshape of the pulse that is applied to the indicator versus the associated noise level in the receiver output. Good pulse waveshape requires wide-band response, whereas low noise output requires narrow-band response. The problem of compromising between these contradictory factors is investigated by means of a receiver that has a fixed signal-to-noise ratio at its mixer, and adjustable bandwidth. In order to receive and to reproduce satisfactorily rectangular pulses equal in amplitude to the noise voltage, the receiver bandwidth should be equal to 2 divided by the pulse width in seconds; the answer is in Hz units. From practical considerations, designers generally add 1 MHz to the calculated bandwidth to allow for residual frequency drift in both the transmitter and in the local oscillator.

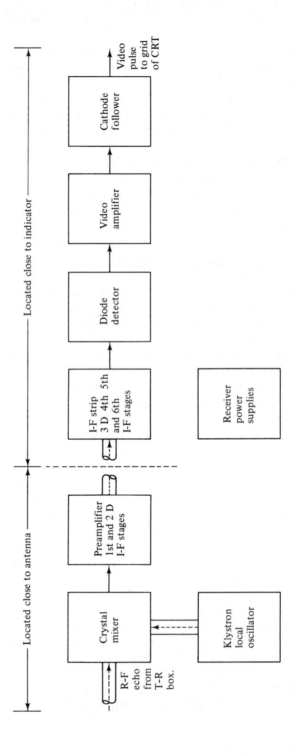

Figure 4-14 Block diagram of a basic radar receiver.

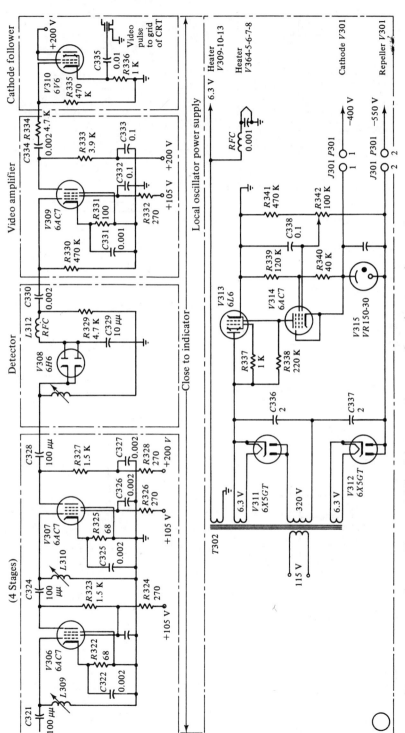

Figure 4-15 Schematic diagram for a basic radar receiver.

105

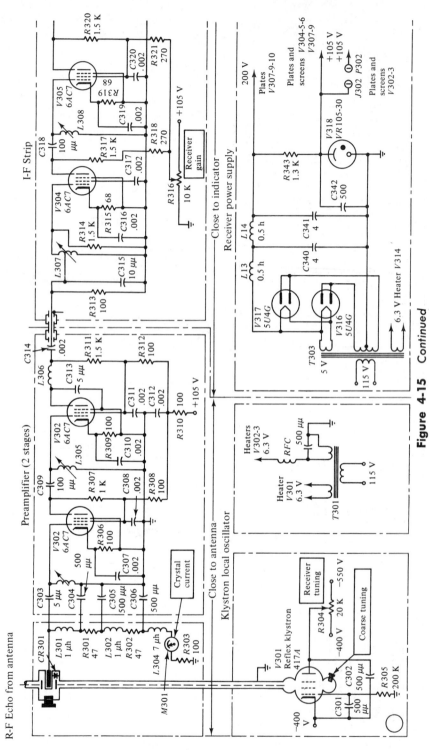

Figure 4-15 Continued

With reference to Fig. 4-15, the klystron applies a frequency to the crystal mixer 30 MHz below the incoming signal frequency. Rough adjustment of the klystron frequency is obtained by tuning the resonant cavity with the threaded plug. Fine tuning over the desired portion of the frequency range is obtained by adjusting the repeller voltage with a tuning potentiometer R304. Injection voltage into the mixer normally provides a crystal current of approximately 0.3 mA. The mixer is the chief source of noise, unless it is preceded by RF amplifiers. When RF amplification is utilized, the RF amplifier becomes the chief source of noise. Gain of the silicon crystal mixer is less than unity; however, its noise output is less than that of other types of mixers. The signals to be heterodyned are applied to the crystal mixer by a resonant cavity. A resonant cavity has comparatively low losses.

An equivalent circuit diagram for the exemplified crystal mixer is shown in Fig. 4-16. The mixing cavity is represented as the inductor L, tuned to the signal frequency by the capacitor C_p. Physically, C_p is a screw plug in the side of the resonant cavity. *Output from the local oscillator differs in frequency from the incoming signal by 30 MHz.*

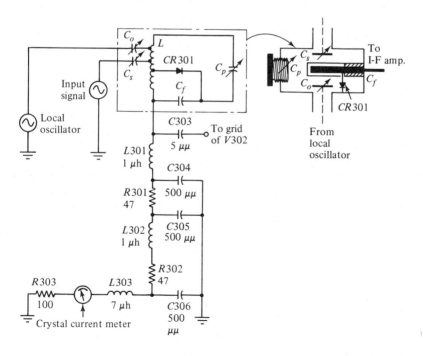

Figure 4-16 Equivalent circuit diagram for the crystal mixer.

This output and the received signal are combined in the mixer to produce a current that contains several frequency components. These frequencies include the frequencies of the incoming and local-oscillator signals, their higher harmonics, and their sum and difference. This difference frequency is selected as the 30-MHz IF frequency and is fed to the first IF amplifier. Probe C_o couples the local-oscillator signal into the mixer cavity and is adjusted so that the crystal current is approximately 0.3 mA. Final matching is accomplished by adjustment of C_p, C_s, the probe that couples the signal input from the antenna into the mixer cavity. C_s is adjusted so that maximum echo signal is injected into the cavity.

Next, the IF signal generated in the crystal is brought out by a lead from the base of the cavity. High-frequency components are bypassed by filter capacitor C_f, which is built into the base of the cavity. Circuit elements L301, C304, R301, C305, L302, R302, and C306 form a three-section L-type filter, and operate to filter the IF frequency out of the circuit of the crystal-current meter. The IF section of the receiver includes six amplifier stages. The first and second stages are located with the mixer and local oscillator in the immediate vicinity of the antenna. In turn, the remaining stages are located in the receiver chassis along with the detector, video amplifier, and cathode follower. All IF stages are adjusted for maximum response at a center frequency of 30 MHz, and the amplifier response is broadened to provide an overall bandpass of 2 MHz.

A simplified diagram of the IF input circuit is shown in Fig. 4-17. Inductor L304 is tuned to resonate at 30 MHz. The crystal mixer can be regarded as a resistor shunted across capacitor C_f. Its reflected resistance that loads the tuned circuit is somewhat greater because of the step-up ratio of C_f to C303. Thus, the voltage applied to the grid of V302 is the crystal voltage stepped up several times. The two preamplifiers are conventional. Their interstage coupling network is a resonant circuit loaded by resistor R307 to broaden the response. The output circuit matches the low input impedance of the coaxial transmission line to the plate of the second IF tube. This circuit consists of the output capacitance of the tube in parallel with C313, inductor L306, and resistor R311, shunted by the impedance of the coaxial line. These elements can be considered as a resonant step-down autotransformer. In turn, the transmission line transports the IF signal to the balance of the receiver, and is terminated in its proper load by R313 and the tapped inductor L307.

For shielding purposes, the last four IF stages are enclosed in a metal box. Since the tuning of the IF coils is not unduly critical, tube replacement does not necessitate realignment of the tuned loads. Re-

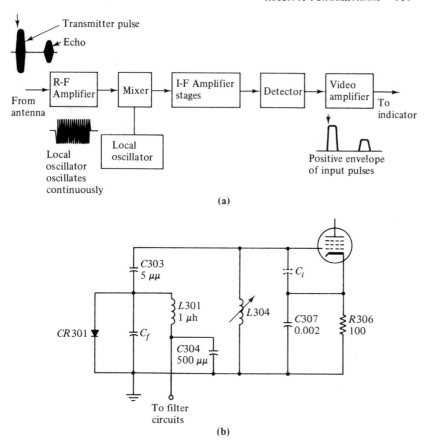

Figure 4-17 Schematic diagram for the exemplified IF input stage. (**a**) Signal relations in receiver subsystem; (**b**) simplified IF configuration.

ceiver gain is controlled by R316, which regulates the plate and screen voltages of the third and fourth IF stages. IF signals are converted to video-frequency signals by a detector diode. The IF potential across L311 is applied between the diode and ground. In turn, the video current that follows the envelope of the IF signal flows through V308, choke coil L312, and load resistor R329. Capacitor C329 bypasses the IF component of the video output signal from the detector. Thus, the video signal is developed as a negative pulse that drops chiefly across R329; it is then applied to the video amplifier via C330.

In addition to providing filter action, L312 and C329 function as a high-frequency compensating arrangement to improve the frequency response of the video amplifier. *The midfrequency gain of the stage is approximately 35 times, and the bandwidth between half-power points is 0.7MHz.* The positive amplified video signals are developed across R333 and are applied to the cathode-follower input circuit via C334. Resistor R334 tends to prevent strong signals on the grid of V310 from charging C334. Video pulses are fed to the indicator grid through a cathode-follower stage. A maximum video signal of about +90 volts on the input of the cathode follower causes its plate current to increase from 20 mA to approximately 92 mA, thereby producing an output of 72 volts and reducing the bias to 2 volts. An output impedance of approximately 200 ohms is realized, which effectively minimizes pickup by the oscilloscope lead and reduces interstage coupling between the oscilloscope input circuit and the receiver circuits. In addition, the loading effect on the video amplifier is negligible, because of the high input impedance of the cathode-follower stage.

5

Basic
Telemetry Systems

Telemetry is defined as the science of sensing and measuring physical information at some remote location and transmitting the data to a convenient location to be read and recorded. *Transmission may be in the form of either analog or digital signal modes.* A telemeter is a complete measuring, transmitting, and receiving apparatus for indicating, recording, or integrating the value of a physical quantity at a distance by electric or electromagnetic wave means. The Atmospheric Structure satellite was the first scientific Earth satellite containing a pulse-code modulation (PCM) telemetry system; it was a frequency-modulated (FM) pulse-code system. Previous satellites utilized a pulse-frequency modulation (PFM) system; these were amplitude-modulated (AM) systems. The solid-state PCM system provided an output power of 500 mW, and had a capability of 40 separate channels of information in digital form.

Examples of various forms of pulse modulation are shown in Fig. 5-1. Pulse-amplitude modulation (PAM) uses the modulating wave to amplitude-modulate a pulse carrier. Pulse-duration modulation (PDM) is also termed pulse-width modulation and pulse-length modulation. PDM is a form of pulse-time modulation in which the duration of a pulse is varied by the modulating wave. Pulse-position modulation (PPM) is another form of pulse-time modulation in which the value of each instantaneous sample of the wave modulates the position in time of a pulse. Pulse-code modulation (PCM) functions by sampling the signal periodically; each sample is quantized and is transmitted in digital

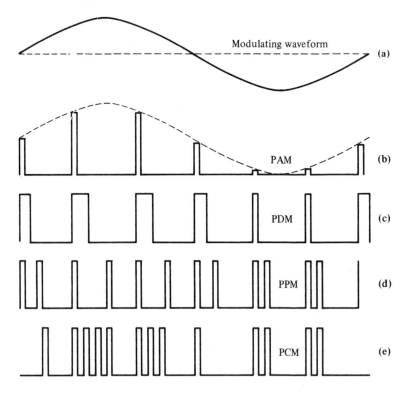

Figure 5-1   Examples of various forms of pulse modulation. **(a)** Modulating sine wave; **(b)** pulse-amplitude modulation; **(c)** pulse-duration modulation; **(d)** pulse-position modulation; **(e)** pulse-code modulation.

binary code. In a PCM system, one complete sampling of words or channels at a given rate is called a *telemetry frame*. The telemetry frame rate is the frequency corresponding to the period of one frame. Values of units such as pressure, radiation intensity, speed, temperature, and so on may be transferred. Telemetry links may consist of cables, electromagnetic waves, power lines, or a combination of such means. Comparatively sophisticated encoding, multiplexing, and decoding techniques are in general use. *Most systems employ some form of pulse modulation.*

5-2   SYSTEM OVERVIEW

Telemetry data are generated by a transducer, which usually operates in an analog mode. Data are indicated and/or recorded by various instruments, such as an oscillograph or X-Y recorder. An X-Y recorder provides a permanent record on a moving strip of paper. The basic plan of a typical telemetry system is shown in Fig. 5-2. Observe that an electromechanical transducer is included at the transmitting location. This transducer converts the physical property to be measured into an electrical signal that modulates a radio-frequency carrier wave. In turn, electromagnetic waves are radiated from the transmitting antenna. These waves are intercepted by the receiving antenna, processed, and applied to a recorder. In this example, an X-Y plotter provides a permanent record. If both the receiver and the transmitter are land-based, a cable network may alternatively be employed for transport of the electrical signals.

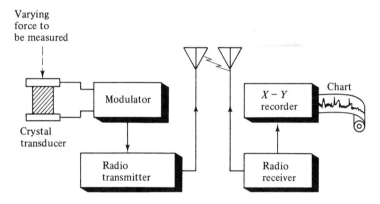

Figure 5-2 Simplified block diagram for a typical telemetry system.

In addition to the basic elements shown in Fig. 5-2, *an information processor called a decoder is usually required, inasmuch as more than one information channel is ordinarily modulated on the same carrier.* Accordingly, each information channel is separately encoded into the transmitted carrier. In turn, the receiver includes a decoder that separates the information in each channel and directs these separate trains of information to individual indicators or recorders. Consider next the skeleton block diagram for a three-channel telemetry system depicted in Fig. 5-3. Note that three independent forces are being measured and all three values are being transmitted on a single RF carrier. To avoid interference among these three signals, some means of encoding and decoding is necessarily provided for each channel. *Some designers prefer to assign an individual subcarrier source to each channel,* with the measured force modulated upon its particular subcarrier.

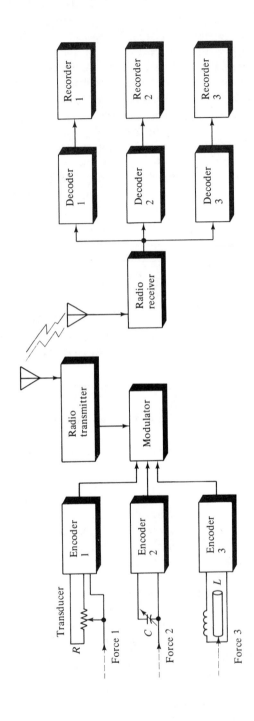

Figure 5-3 Skeleton block diagram for a three-channel telemetry system.

Thus, encoder 1 might operate at a subcarrier frequency of 1.3 kHz, encoder 2 at 1.7 kHz, and encoder 3 at 2.3 kHz.

Next, at the receiver location, decoder 1 will contain a bandpass filter with a center frequency of 1.3 kHz, decoder 2 a center frequency of 1.7 kHz, and decoder 3 a center frequency of 2.3 kHz. After the decoding process, the demodulated outputs are applied to recorders 1, 2, and 3 respectively. Various modes of modulation may be employed. For example, the measured force may amplitude-modulate its subcarrier, or it may frequency-modulate its subcarrier. Again, pulse-duration modulation, or pulse-code modulation may be utilized. Note that in the arrangement of Fig. 5-3, the transducers may frequency-modulate the encoder oscillators, and in turn the frequency-modulated oscillators may simultaneously frequency-modulate the radio transmitter.

The foregoing design approach is called a frequency-sharing system. Each information channel is represented by a certain band of frequencies. An alternative term is a frequency-division multiplex system. The transmitter generates a complex composite signal. Observe that a frequency-division multiplex system may be of FM-AM design, instead of the FM-FM design that was described above. That is, in an FM-AM telemetry system, the transducers frequency-modulate their respective subcarrier sources, and the combined FM signal is amplitude-modulated on the RF carrier. In this arrangement, the receivers utilize AM circuitry followed by FM decoders. In PAM transmission, the received signal is rectified and then integrated to recover its modulation envelope, as depicted in Fig. 5-4. A sampling process is utilized at the transmitter, followed by a reconstitution process at the receiver.

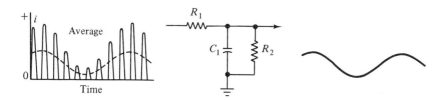

Figure 5-4   Integration recovers the modulation envelope from a rectified PAM signal.

PCM is characterized as a *quantizing* system, wherein the modulating waveform is sampled; each sample must assume a nearest whole value, such as 0.1, 0.2, 0.3, and so on, with respect to the maximum transmitted value. Consequently, the quantized pulses have values that are coded. *Binary pulse-code modulation* is a type of PCM that is based on a code for each sample in terms of pulses and spaces. Pulses are either "on" or "off," and the amplitude of the sample corresponds to the number of "on" pulses that are transmitted within a given interval. *Ternary PCM is also used by some designers.* This is a telemetry system that transmits both positive and negative pulses with spaced intervals. Each type of pulse modulation requires appropriate processing equipment at the receiver location. Binary PCM has an advantage in some situations in that it is suitable for direct processing by digital computers.

A common feature of all pulse-modulation systems is the principle of sampling. This technique permits transmission of the essential modulating information without continuous modulation of the RF carrier. In turn, the "off" intervals between pulses can be employed for transmitting additional information. Thus, transmitter efficiency is increased; however, there is a trade-off involved, in that the system bandwidth must be increased when an additional channel is employed. When constant-amplitude pulses are utilized in a telemetry system, a limiter is included at the receiver to minimize noise voltages in the output signal. In addition to permitting the use of limiters, *constant-amplitude pulses provide another design advantage, in that the system amplifiers do not have to be carefully linearized.* It is evident that even if the amplifiers are quite nonlinear, precisely the same information is transferred by means of constant-amplitude pulses. Information transfer by pulse-modulation methods is comparable to telegraphy system operation, except that the pulse repetition rate is much faster. Since comparatively large bandwidths are required by telemetry systems, carrier frequencies are generally allocated in the super-high-frequency ranges.

5-3   SIGNAL PROCESSING

Aliasing is defined as the introduction of error into the Fourier analysis of a discrete sampling of continuous data when components with frequencies too great to be analyzed with the sampling interval being used contribute to the amplitudes of lower-frequency components. In other words, severe distortion in reconstitution of the modulating wave, owing to an inadequate sampling rate, can result in excessive system error. Accordingly, system designers choose a sampling rate that is sufficiently great that objectionable distortion and aliasing in the reconstitution process is avoided. The general requirements are shown in Fig. 5-5. When the sampling rate is rapid, this factor eliminates the possibility of excessive error at the receiver output. On the other hand, a slow sam-

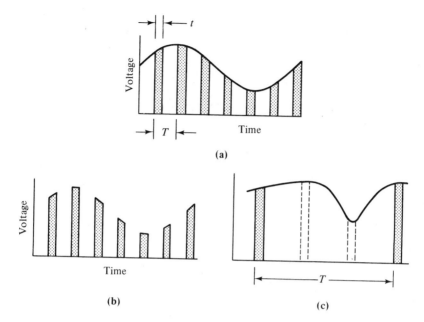

Figure 5-5 Telemetry sampling process. **(a)** Comparatively rapid sampling; **(b)** receiver output; **(c)** distortion (aliasing) results from low-rate sampling.

pling rate may result in the loss of essential modulating waveform detail, with resultant aliasing.

With reference to Fig. 5-5(a), t denotes the sampling pulse width, and T indicates the pulse-repetition rate. The system designer often wishes to make the "off" time between pulses comparatively large. In turn, the value of t is made as small as possible. Similarly, the value of T is made as large as possible. However, extremely narrow pulses contain comparatively little energy, and they may be unable to compete effectively with the prevailing noise level. Similarly, excessively large values of T result in aliasing difficulties. In turn, the designer must make a judicious tradeoff in specifying these parameters. *Pulse multiplexing* techniques exploit "off" times between pulses in a series to insert another series of pulses and thereby provide an additional information channel. The receiver is gated to separate one pulse train from another. Thus, pulse multiplexing is basically a simple system parameter.

A square wave and a sawtooth wave are being sampled in the example of Fig. 5-6; the square wave is sampled by one series of pulses, and the sawtooth is sampled by another series of pulses. Evidently, if the receiver is gated in synchronism, the output from one gate will be

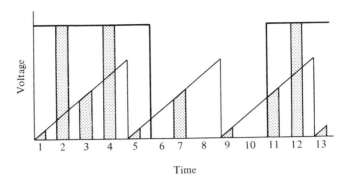

Figure 5-6 Square wave and pulse waveforms being sampled by two pulse trains.

the sampled square wave, and the output from the other gate will be the sampled sawtooth wave. Designers use a simple equation to calculate the minimum pulse repetition rate that should be used:

$$\text{Repetition rate} = 2f_m$$

where f_m is the bandwidth of the modulating waveform. Thus, if the modulating waveform has a bandwidth of 400 Hz, then the pulse sampling rate should be at least 800 pulses per second (pps). Note that the pulses may be made as narrow as desired, within the capability of the transmitter and receiver to process the pulses and to reconstitute the modulating waveforms without excessive noise interference. *The designer may multiplex any number of signal waveforms, provided that ample system capacity is available.* Multiplexing of four signal waveforms is depicted in Fig. 5-7. Pulses marked "1" sample one waveform; pulses marked "2" sample another waveform, and so on. Accordingly, four gated channels are provided at the receiver; the "1" channel is

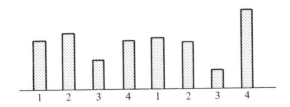

Figure 5-7 Example of four pulse-multiplexed channels.

gated on each time that the pulse marked "1" occurs. Of course, the "1" channel is gated off during the time that the pulse marked "2" occurs; during this time the "2" channel is gated on.

5-4 PULSE CHARACTERISTICS

A telemetry system requires a bandwidth that depends on the rise time or the fall time of the pulse, whichever is the faster (see Fig. 5-8). This bandwidth is stated by the equation

$$BW = \frac{1}{2t}$$

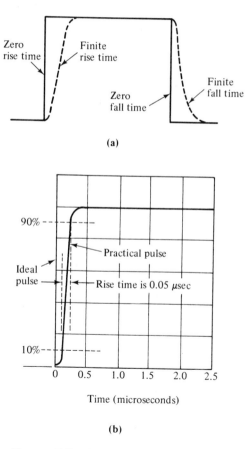

(a)

(b)

Figure 5-8 Rise-time characteristics.

where t is equal to the rise (or fall) time of the pulse. As an illustration, if the pulse has a rise time that is faster than its fall time, and its rise time is equal to 0.05 μs, the system bandwidth should be at least 10 MHz. *The system designer must also take measures to minimize interchannel crosstalk.* In other words, pulses in one channel should not produce objectionable interference in another channel. This difficulty is avoided by utilizing a sufficient time interval between successive pulses. As a general rule, it is necessary that the last pulse in a given channel be allowed time to decay to at least 50 percent of maximum amplitude before the first pulse in the next channel starts to rise. To avoid objectionable interference in a multiplexed PAM system, an additional amount of decay time should be observed.

When pulses are amplitude-modulated on a carrier, upper and lower sidebands are generated. In turn, twice as much RF bandwidth must be employed than is stipulated by the foregoing basic equation. That is, the RF bandwidth is given by

$$BW_{RF} = 1/t$$

where t is equal to the rise (or fall) time of the pulse, whichever is faster. In the case of pulse frequency modulation (PFM), the required bandwidth depends on the modulation index of the frequency-modulated carrier:

$$BW_{RF} = \frac{m + 1}{t}$$

where m is the modulation index. As an illustration, if the pulse rise time is 0.05 μs, and the modulation index is equal to 0.1, the RF band-

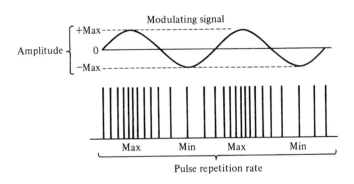

Figure 5-9   Example of pulse frequency modulation.

width should be at least 22 MHz. An example of pulse frequency modulation is depicted in Fig. 5-9. To accommodate wide-band transmissions, radio telemetry systems operate in high-frequency, very-high frequency (VHF), and ultra-high frequency (UHF) regions of the electromagnetic wave spectrum (see Table 5-1). A standardized RF telemetry band extends from 216 to 235 MHz. Frequencies as low as 20 MHz have been used in space research. Another example of a standardized telemetry band is 2200–2300 MHz. Subcarrier frequencies range typically from 400 Hz to 70 kHz.

TABLE 5-1

Basic Frequency Bands

Band	Frequency Spectrum (kHz)*	Wavelength in Meters*
VLF (very low)	10–30	30,000–10,000
LF (low)	30–300	10,000–1,000
MF (medium)	300–3,000	1,000–100
HF (high)	3,000–30,000	100–10
VHF (very high)	30,000–300,000	10–1
UHF (ultra high)	300,000–3,000,000	1–0.1
SHF (super high)	3,000,000–30,000,000	0.1–0.01
EHF (extremely high)	30,000,000–300,000,000	0.01–0.001

* 1 meter = 39.37 inches
  1 kilohertz = 1000 hertz
  1 megahertz = 1000 kilohertz
  1 gigahertz = 1000 megahertz

5-5 SENSORS AND SIGNAL CONDITIONERS

A wide variety of sensors (transducers) is utilized in telemetry systems. Physical properties include frequency, strain, flow rate and/or quantity, gas density, mechanical position, electrical current, magnetic field intensity, radiation intensity, impact force, voltage, acceleration, vibration amplitude and/or frequency, motional velocity and/or direction, liquid level, temperature, pressure, and so on. A typical strain gauge for actuating three channels is depicted in Fig. 5-10. As shown in Fig. 5-11, a sensor or transducer is generally followed by a signal conditioner. These signal conditioners are grouped into high-level and low-level types. A few systems do not require signal conditioning.

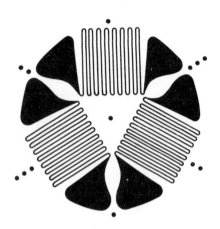

Figure 5-10   A three-channel strain-gauge arrangement.

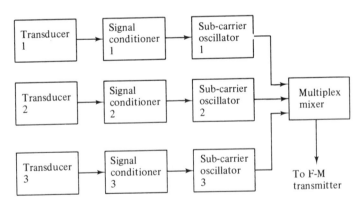

Figure 5-11   Example of AM-FM telemetry transmitter driver with signal conditioners.

It is apparent that many kinds of sensors cannot operate a telemetry system directly. As an illustration, the output from a quartz-crystal transducer has a very low level. On the other hand, *a typical telemetry system requires an input level up to 5 volts.* Consequently, the signal conditioner following a quartz-crystal transducer must provide substantial amplification. Other kinds of sensors have adequate output voltage, with a poor impedance match. A typical telemetry system has an input impedance of 100,000 ohms. In turn, if the sensor has a widely different impedance value, the signal conditioner will be designed for appropriate impedance transformation. Some kinds of sensors have an AC output; proximity devices are examples. Thus, to minimize the required channel bandwidth, the signal conditioner is designed to change the FM AC output from the sensor to a varying DC voltage.

In the foregoing example, the incoming signal is processed by an FM receiver, as shown in Fig. 5-12. In turn, the output from the receiver is a composite signal comprising the modulated subcarriers. It is a multiplexed signal that can be processed, or stored and processed at a later time. For example, the output from the receiver can be applied to a tape recorder. Subcarrier discriminators are employed for decoding the composite signal. Note that the patch panel indicated in Fig. 5-12 comprises a switching facility that permits the information from a channel to be viewed directly, as on an oscilloscope screen or meter scale, or to be recorded with an X-Y plotter. When channel information is monitored directly, the process is called *real-time monitoring.* This term denotes that the information is read out simultaneously with its reception.

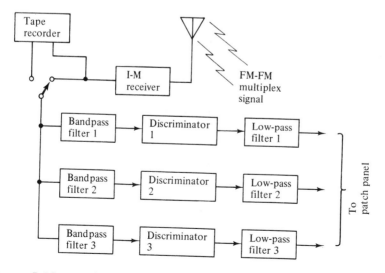

Figure 5-12 Patch panel permits use of a tape recorder for data storage.

5-6 ANALOG-DIGITAL CONVERSION

Telemetry systems make extensive use of analog-digital (A&D) converters. This device transforms analog data into digitally coded data. An A&D converter is also called an encoder, an ADC device, or a digitizer. Consider a PCM transmitter multiplexing arrangement. In the encoding circuits, A&D converters translate a continuously changing signal into digital form. In turn, the encoder responds by producing a series of pulses that represent the instantaneous value of the signal. A binary coding relation is often employed. Four binary digits (four bits) are utilized in a typical system. Zero voltage corresponds to the binary number 0000. Note that a total of 16 proportional values can be represented in this coding arrangement. A half-of-maximum voltage value corresponds to the binary number 1000, or 8/16 in decimal notation. Suitable circuitry is provided at the receiver to decode the incoming binary signal. Means are provided in both the transmitting and the receiving systems for sampling the various channels at regular intervals. This sampling process is termed *commutation*. It denotes sequential sampling, on a repetitive time-sharing basis, of several signals. Sampling may be followed by transmission and/or recording of the associated channel signals. *Commutation duty cycle* denotes the ratio of the "on" time for a particular channel to the total time assigned to that channel. The total time required to sample all of the channel signals once is called the *commutation frame period*. Electronic switches are commonly used for commutation. The receiver utilizes an inverse device termed a *decommutator*.

5-7 TYPICAL PPM TELEMETRY SYSTEM

A representative PPM telemetry system utilizes eight channels. These channels are multiplexed, and operate with the composite pulse waveform depicted in Fig. 5-13. In order to identify individual channels at the receiver, an identifying pulse is required between the channel 8 and the channel 1 pulse. This pulse is also called a marker pulse. It is comparable to the vertical synchronizing pulse in a composite television signal. Because the marker pulse is wider than the code pulses, it can be separated from the composite telemetry signal by suitable filter circuitry. Some telemetry systems use other forms of marker pulses. For example, a pair of code pulses with a special spacing may be employed. Other system designers utilize a code pulse as a marker pulse by means of frequency shift keying (FSK). That is, the marker pulse

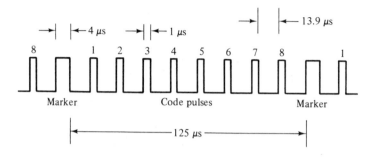

Figure 5-13 Composite telemetry signal for eight PPM multiplexed channels.

frequency-modulates the RF carrier, whereas the code pulses amplitude-modulate the carrier.

A code pulse may occupy approximately 10 percent of the width over the interval from one marker to the next in a PPM composite signal. This interval is a function of the number of multiplexed channels. Note that a typical marker pulse has a width of 1 μs and a repetition rate of 125 μs. As shown in Fig. 5-13, each channel shares the available time equally. In this example, PPM action can displace each pulse a maximum of ± 6 μs from its quiescent or unmodulated position. There is an elapsed time of 125 μs between successive marker pulses, and because eight channels are multiplexed, each channel has a width of approximately 15 μs. As noted before, the total maximum displacement of each code pulse is 12 μs. Thus, a duty cycle of approximately 10 is utilized; the duty cycle is equal to the pulse width divided by the pulse repetition period, as shown in Fig. 5-14.

Note that the sampling rate shown in Fig. 5-13 provides for trans-

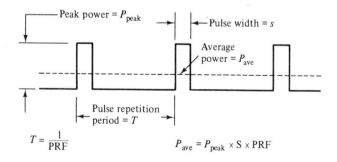

Figure 5-14 Example of a 1/7 duty cycle in a pulse train.

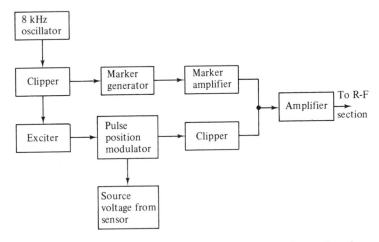

Figure 5-15 Arrangement for generating the code pulse for one channel, and the marker pulse.

mission of waveforms with frequency components up to 3 kHz. A block diagram of the circuit sections used to modulate the pulses for one channel is depicted in Fig. 5-15. The marker pulse has a width of 4 μs and serves both as an identification pulse for the receiver and as a master timing pulse at the transmitter. In this example, the repetition rate of the marker pulse is 8 kHz. A rectangular wave output is provided by the 8-kHz oscillator, as seen in Fig. 5-16. This is an AC waveform with an average value of zero. To improve its rise time, a clipper stage is utilized between the 8-kHz oscillator and the marker generator (Fig. 5-15). The output from the clipper is also applied to the exciter circuit. In the marker generator, a pulse is produced with a

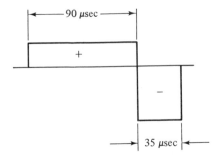

Figure 5-16 Rectangular output waveform from the 8-kHz oscillator.

width of 4 μs. The timing of the marker pulse is controlled by the wave-form in Fig. 5-16. In the exciter section, a code pulse is produced that has a width of 2 μs. This code pulse is applied in turn to the pulse-position modulator section. Each time that the amplitude of the output waveform from the exciter reaches 15 V, a code pulse is generated. Note that a varying modulating voltage is provided from the sensor.

Because the exciter waveform does not rise instantaneously, but has a leading edge that slopes, the trigger point on its leading edge is determined by the height that the sensor voltage lifts the leading edge above zero, as shown in Fig. 5-17. If the sensor output voltage is zero, the pulse-position modulator is triggered at its 50 percent level on its leading edge. However, when the sensor voltage rises to almost its maximum positive value, the pulse-position modulator is triggered near the start of its leading edge. If the sensor output voltage has its maximum positive value, the code pulse is advanced 6 μs from its quiescent or unmodulated position. On the other hand, when the sensor output voltage is near its maximum negative value, the pulse-position modulator is triggered near the top of its leading edge. Accordingly, the code pulse is delayed 6 μs from its unmodulated position. With reference to Fig. 5-15, the code pulses are clipped and are then mixed with the marker pulses. This composite waveform is amplified and clipped again before it is applied to the RF section. Thereby, the rise time of the pulses is improved.

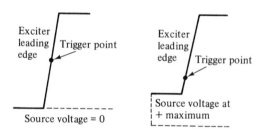

Figure 5-17 A code pulse is triggered earlier by application of a positive sensor voltage.

Observe next how the pulses are processed for radiation from the telemetry transmitter in Fig. 5-18. The pulses are amplified and clipped by two video-frequency amplifier stages and are then applied to the modulator section. This modulator functions also as an amplifier, and raises the pulse amplitude to 1 kV. In turn, negative-polarity pulses are applied to the klystron oscillator; amplitude modulation is em-

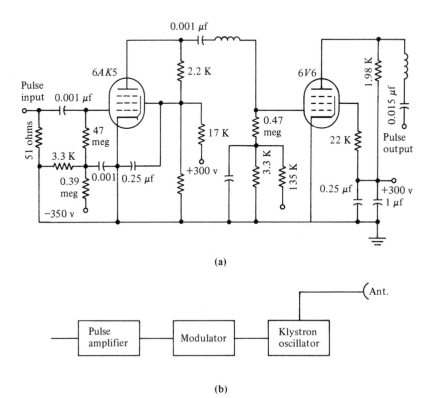

(a)

(b)

Figure 5-18 Block diagram of the RF system, and skeleton circuit of the pulse-amplifier section. **(a)** Pulse-amplifier configuration; **(b)** klystron oscillator is amplitude-modulated.

ployed. The klystron is cathode-driven and generates a UHF output of approximately ¼ W. At the receiver, code pulses from the eight channels are separated and fed to eight outputs. Proportions of the incoming composite telemetry signal are indicated in Fig. 5-19. Electronic switching is utilized to separate the code pulses for each channel. Gate pulses are timed to coincide with an associated code pulse that is to be gated out of the train. Gated-out pulses are then amplified and fed to an integrator for reconstitution of the envelope waveform originated by the sensor. In turn, the integrator output may be displayed on an oscilloscope screen, or a permanent record may be made by an X-Y recorder. The block diagram shown in Fig. 5-19(b) is for a single decoder; eight similar decoders are utilized in the system. The only difference between one decoder and another is in the timing of the gating pulse.

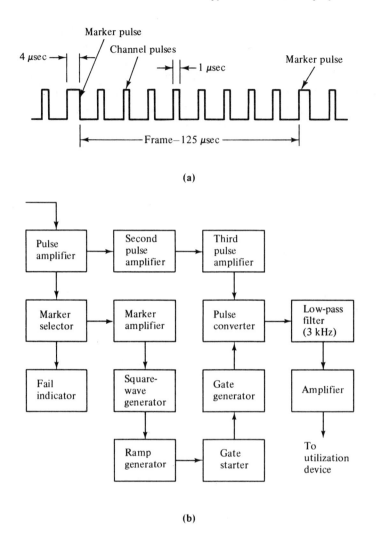

Figure 5-19 Typical one-channel telemetry decoder arrangement.
(a) Pulse train input waveform; (b) decoder block diagram.

The incoming composite telemetry signal is amplified and fed to a marker selector, as shown in Fig. 5-19. This pulse amplitude is normally 10 volts, approximately. Negative polarity pulses are supplied by the first pulse amplifier. *The system designer may employ either the leading edge or the trailing edge of a pulse to trigger the decoder circuit.* Trailing-edge triggering is employed in this example, because the fall time is faster than the rise time, and because the trailing edge has a better signal-to-noise ratio. Note that a more precise trigger action is provided by a steeper pulse edge. Each time that the second pulse amplifier in Fig. 5-19 is triggered, it generates a pulse in response. This is a *regenerated pulse;* it has a better output waveform than that of the incoming pulse, particularly when the receiver is operated at a great distance from the telemetry transmitter. As would be anticipated, the regenerated pulses are slightly delayed with respect to the incoming pulses.

Positive pulses with an amplitude of approximately 25 volts are provided by the third pulse amplifier in Fig. 5-19. Note that negative 10-volt pulses are applied to the marker-selector section from the first pulse amplifier. This marker-selector section includes an integrator with a time constant such that the 1-μs code pulses do not drive the grid of the subsequent tube appreciably. However, the 4-μs marker pulses are passed to the grid and drive the tube into cutoff. Accordingly, an output pulse is generated by the marker-selector section. This output pulse is applied to a marker amplifier that functions to square the marker pulse, thereby forming a precise trigger pulse for application to the square-wave generator.

Observe that the square-wave generator indicated in Fig. 5-19 will have a repetition rate of 8 kHz, inasmuch as it is synchronized by the marker pulse. In turn, the following ramp (sawtooth) generator converts the square wave into a sawtooth wave, which in turn is applied to the gate-starter section. This gate-starter section employs an amplifier that is biased beyond cutoff; in turn, the ramp waveform is ineffective until the time that the intended code pulse appears. Thereupon, the gate-starter section develops an output that triggers the gate generator. This gate generator is basically a one-shot multivibrator that produces a 13 μs gate pulse; note that this gate pulse is coincident with the code pulse that is to be separated from the composite telemetry signal. Separation is accomplished by elevation and slicing (see Fig. 5-20).

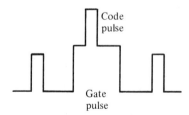

Figure 5-20 A code pulse is elevated to permit separation.

Thus, output from the gate generator is applied to the pulse converter, thereby doubling the effective amplitude of the code pulse. In turn, the pulse-amplitude tube is biased so that only the elevated code pulse exceeds the conduction threshold. A code pulse may occupy any 1 μs interval within the 13 μs gate pulse period.

Recall that *the position of a code pulse within the gating interval is a measure of the sensor voltage at the instant of sampling.* Accordingly, the position of the pulse must be translated into a corresponding voltage amplitude. This function is accomplished by the pulse-converter stage. In other words, a bistable multivibrator, or flip-flop, is triggered by the leading edge of the code pulse and is turned off by the trailing edge of the gate pulse. It is evident that the output from the pulse converter is another pulse that has a width proportional to the position of the code pulse. This width is in the range from 1 to 12 μs. In turn, the output pulses from the pulse converter are processed by a 3-kHz low-pass filter that functions essentially as an integrator. That is, its output voltage is proportional to the width of the converter output pulses. Accordingly, the output from the integrator reproduces the waveform of the sensor voltage.

5-8   PDM TELEMETRY SYSTEM

Various other modes of multiplex telemetry pulse communication are used by system designers. One basic mode employs pulse-width multiplexing. Note that pulses do not recur at fixed intervals in a PDM telemetry system. As seen in Fig. 5-21, the code pulses in each channel may follow one another in a dense series in which they are separated by a reference time *t.* Next, the code pulses may have a less dense distribution in which they are separated by a reference time of 2*t,* or 3*t.* The diagram depicts code pulses that start with dense distribution, followed by code pulses with progressively less dense distribution. Information is transferred as a function of pulse-repetition rate in this arrangement.

Observe in the waveform of Fig. 5-21 that there is space available between successive code pulses for the multiplexing of additional pulses, even in the region of maximum density. Also shown in Fig. 5-21 is a block diagram for a system that employs three multiplexed channels. Since a different pulse width is utilized in each channel, a channel pulse can also serve as a marker pulse and thereby identify its own channel. *This particular design is called a pulse-width coding system.* Note that when an information pulse is fed to the channel 1 pulse modulator, a channel pulse is generated that has comparatively narrow width, such as 30 μs. Again, the channel 2 pulse modulator may produce a channel

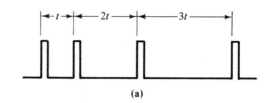

(a)

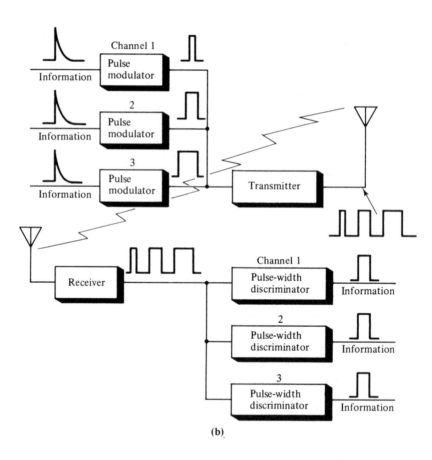

(b)

Figure 5-21 Typical pulse-width multiplex telemetry system for three channels. **(a)** Information is transferred by the pulse repetition rate; **(b)** channel identification is accomplished by pulse widths.

pulse with a width of 60 μs, and the channel 3 pulse modulator may develop a channel pulse with a width of 90 μs.

In the receiver system depicted in Fig. 5-21, *the pulse-width dis-*

criminators are designed to accept or reject a code pulse on the basis of its width. That is, an output pulse is obtained from a discriminator only in the event that the applied pulse has a specified width. On the other hand, output pulses from the discriminators all have the same width. Refer to Fig. 5-22. This is a typical pulse-width discriminator block diagram. Inasmuch as the incoming pulse (1) is likely to be contaminated by noise voltages, the signal is first passed through a limiter or slicer (top-and-bottom clipper). This processing greatly improves the signal-to-noise ratio. Output signals from the slicer are applied to an inverter that reverses the pulse polarity from (2) to (3). Observe that the noninverted pulses are differentiated at (4).

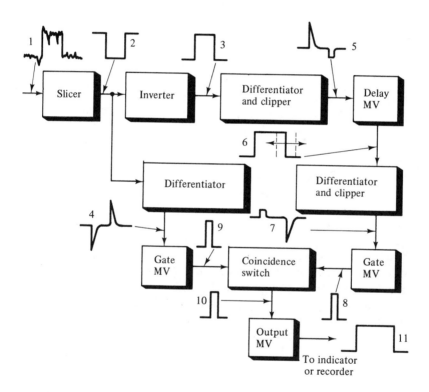

Figure 5-22 Typical pulse-width discriminator block diagram.

Proceed to step (9); here, a one-shot multivibrator produces a gating pulse of precise width that is applied to the coincidence switch tube. Note also that this gate multivibrator has the additional function of delaying the gating pulse by a time period equal to the pulse width.

In other words, the gate multivibrator is triggered by the positive-going edge of the differentiated waveform at (4). Simultaneously, the inverter output has been differentiated at (5); this differentiated waveform then triggers a delay multivibrator. Potentiometer control is provided for precise delay adjustment. Thus, the trailing edge of (6) can be adjusted as required by the operator for optimum signal reproduction on a particular channel.

Next, observe that the waveform depicted at (5) in Fig. 5-22 is obtained from a germanium diode; an asymmetrical output waveform results that is advantageous in avoiding spurious responses from the delay multivibrator. In turn, the output pulse produced by the delay multivibrator is differentiated and then clipped to develop the waveform depicted at (7). This negative pulse has the function of triggering a second gate multivibrator in order to generate the pulse waveform shown at (8). This pulse waveform is delayed by a time period that is equal to the width of the pulse that is developed by the delay multivibrator. Note that the two gate pulses (8) and (9) are applied to the coincidence-switch tube; this tube is biased to conduct only when two gate pulses appear simultaneously (the voltage sum exceeds the cutoff level). In other words, *the coincidence switch tube functions as a digital-logic AND gate.* In turn, the output multivibrator generates a pulse if it is triggered by a pulse at (1); if an output pulse (11) is generated, it will have a fixed width.

Observe that it is desirable to top-and-bottom clip the incoming pulse (1) to optimize its signal-to-noise ratio. Therefore, the clipper tube is biased to approximately 14 volts beyond cutoff. The incoming pulse has an amplitude of approximately 30 volts; in turn, the first 14 volts are sliced off. A comparatively low plate voltage is utilized, so that the clipper tube will saturate at a grid potential in excess of 17 volts. This results in slicing off the last 13 volts of the incoming pulse. Consequently, the slicer output consists of a 3-volt center section from the incoming pulse. Differentiating action at (5) is provided by a 50-pF capacitor and a 51-kilohm resistor. The resulting time constant is 2.55 μs. A germanium diode is connected in parallel with the 51-kilohm resistor to largely eliminate the negative excursion of the waveform as depicted at (5). A discriminator must be suitably adjusted for operation on a particular channel. Accordingly, the delay multivibrator is adjusted to develop an output pulse at (6) with a width in the range from 20 to 100 μs. Optimum operation of the coincidence switch tube, or AND gate, is obtained when the gating-pulse width is made as great as possible, while preventing objectionable crosstalk. Thus, the gating pulse is adjustable by the operator over a range from 2 to 12 μs.

6

Elements of
Microprocessor System Design

6-1 GENERAL CONSIDERATIONS

A microprocessor is a large-scale integrated circuit that is equivalent to the central processing unit (CPU) of a large digital computer. Microprocessors are used in large computers, in minicomputers, and in microcomputers. Large computers can be designed as multiprocessing systems; these systems are organized with two or more interconnected computers that perform functionally specialized tasks. That is, a large computer with multiprocessing capability contains two or more microprocessors. In the first analysis, *system reliability involves redundancy*. A fault-tolerant digital system employs duplicate microprocessors, so that if one should fail, the other will be automatically switched into the system. Microprocessors are utilized in minicomputers to perform separate functions that were formerly assigned to a single elaborate CPU. A microcomputer comprises a microprocessor, a control memory, a temporary storage memory, and a master clock, as depicted in Fig. 6-1. This arrangement is termed a minimal general-purpose microcomputer.

6-2 FUNDAMENTAL MICROPROCESSOR CHARACTERISTICS

The Intel 4004 microprocessor was the first arrival on the microprocessor scene, and has remained a significant building block for the digital system designer. It operates from $+5$- and -10-V power supplies and features a four-bit parallel CPU with 46 instructions. The 4004 can directly address 4000 eight-bit instruction words of program memory and 5120 bits of data storage random-access memory (RAM). Up to

16 four-bit input ports and 16 four-bit output ports can also be directly addressed. Sixteen index registers are provided internal to the micro-processor for temporary data storage. This microprocessor operates at clock rates up to approximately 750 kHz. Pin assignments for the 4004 are shown in Fig. 6-2.

Data and control lines for the 4004 are summarized as follows:

D0-D3, bidirectional data bus for handling all address and data communication between the microprocessor and the RAM and ROM chips. $\phi1$-$\phi2$, nonoverlapping clock signals that determine the timing of the microprocessor. Sync, synchronization signal that indicates the beginning of the instruction cycle to the RAM and ROM chips. Reset, a "1" level applied to RESET clears all flag and status flip-flops and forces the program counter to 0. RESET must be applied for 64 clock cycles (eight machine cycles) to completely clear all address and index registers. Test, input; the logic state of TEST can be examined with JCN instruction. CM-ROM, line enables a ROM bank and I/O devices that are connected to the CM-ROM line. CM-RAM0 through CM-RAM3, lines function as bank select signals for the RAM chips in the system.

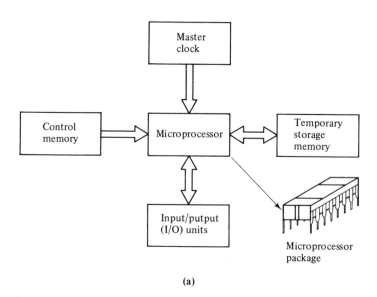

(a)

Figure 6-1 Minimal general-purpose computer functional diagram. **(a)** System block diagram;

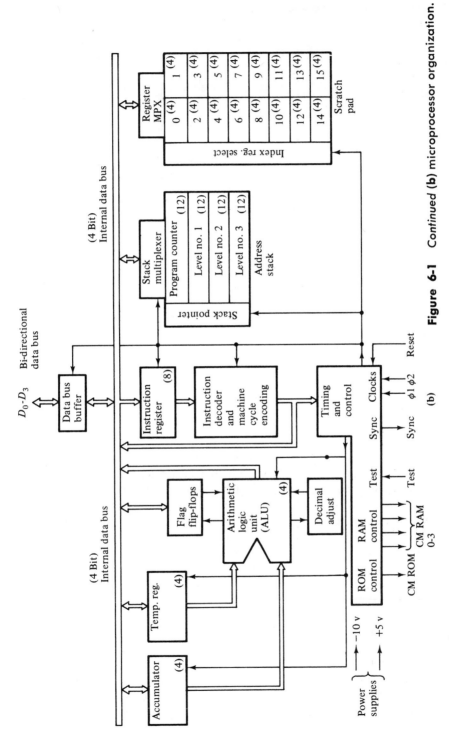

Figure 6-1 Continued **(b)** microprocessor organization.

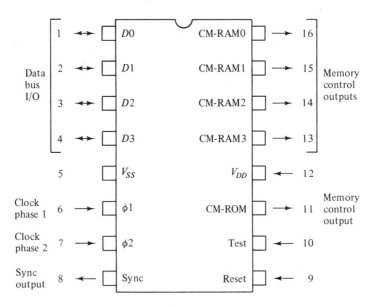

Figure 6-2 Pin assignments for the Intel 4004 microprocessor. (*Courtesy of Hewlett-Packard*)

Probe Connections

Debugging of the microprocessor system during its design phase is greatly facilitated by real-time analysis of program flow, triggering on specific events, and so on. The Hewlett-Packard data domain analyzers illustrated in Fig. 6-3 are particularly well adapted for this task. The 4004 microprocessor does not provide a unique clock for the logic state analyzer at the proper time (end of A3 state) in the instruction cycle. The CM-ROM line is always true at A3 and can be used as a clock signal. However, CM-ROM also occurs at states M2 or X2 during the execution of some instructions. This would result in invalid data's being displayed by the analyzer. By constructing the circuit shown in Fig. 6-4, the designer can ensure a correct state display.

If the portion of the program that the designer wishes to examine is completely contained on one ROM chip, the chip select line (CS) for that ROM can be used as a clock. The probe connections shown in Fig. 6-5 provide a display of the activity on the address line. *A system that will not "come up" can frequently be debugged by monitoring address flow alone.* The 4004 CPU chip has a four-bit data bus, on which the 12-bit address is multiplexed during A1, A2, and A3 states

Figure 6-3 View of the 1607A and 1710B data domain analyzers. (*Courtesy of Hewlett Packard*)

of the 4004 machine cycle. In order to view the demultiplexed 12-bit address on the 1600A, the 4004 system must use 4008/4009 Standard Memory and I/O Interface Set, the 4289 Standard Memory Interface, or similar logic circuits that provide a demultiplexed address bus. If the designer's system uses memory chips that internally decode the multiplexed address, such as the 4001 ROM, the microprocessor data should be monitored as detailed subsequently.

To set the controls, the operator turns the power on and sets the logic state analyzer controls as follows: display mode, Table A; sample mode, SGL. Note that SGL is selected for viewing single-shot events. Press RESET to start the system. The first time that the system passes through the trigger point, the display will be generated and stored. For programs that are looping or cycling through the selected address, select REPET sample mode. The start display is set to ON; trigger mode is set with NORM/ARM to NORM, LOCAL/BUS to LOCAL, and

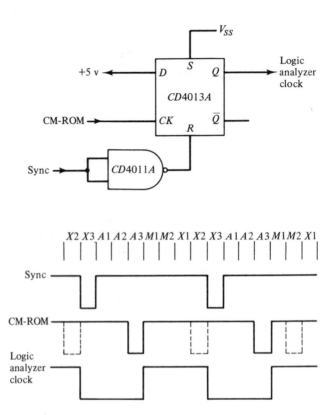

Figure 6-4 Circuit for deriving a clock for the HP 1600A from the 4004 sync and CM-ROM signals. (*Courtesy of Hewlett-Packard*)

OFF/WORD to WORD. The threshold control is set to VAR, and is adjusted to 3.7 V; in the case of TTL compatible systems, the threshold is set to TTL. The LOGIC control is set to POS; CLOCK, to　, and all other pushbuttons are set to their Out position. The display time control is set ccw; qualifiers, OFF; trigger word switches are set to the address at which the operator wishes to trigger; column blanking is set after the display is on-screen and blanking is adjusted to display 12 columns of data.

Next, consider the display interpretation. In this illustration, a segment of a chip tester program for Quad NAND gates is examined. Proper operation is confirmed by a comparison between real-time state analysis shown in Fig. 6-6(a), and the 4004 cross-assembler program listing output depicted in Fig. 6-6(b). The chip tester routine performs the following functions: (1) Sets up bit patterns in the accumu-

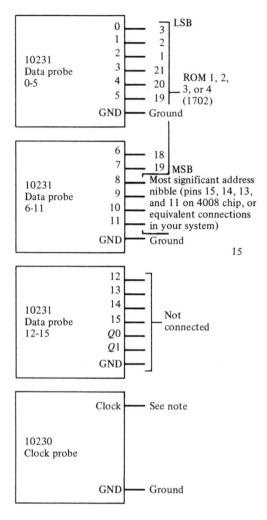

Figure 6-5 Probe connections. (*Courtesy of Hewlett-Packard*)

lator. (2) Outputs the accumulator contents to the NAND gates that are conneced to I/O port 1. (3) Reads the gate outputs. (4) Tests on the gate outputs and indicates whether the chip is good or bad.

Consider the program listing depicted in Fig. 6-6(b). The instructions located in addresses 030 through 033 load the bit pattern 0010 into the accumulator. The next instruction, (WRR), in location 034, writes the accumulator contents to output port 1. The next two instructions (address locations 035 and 036) read the gate outputs present at

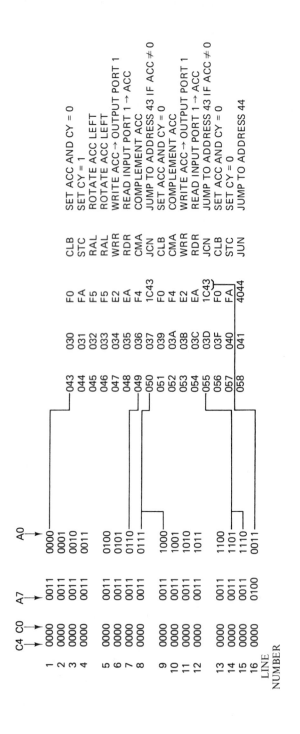

(a)

Figure 6-6 System response to test routine. (a) Real-time state analysis; (b) cross-assembler program listing output. (Courtesy of Hewlett-Packard)

(b)

142

input port 1 into the accumulator and complement the accumulator. Examination of lines 1 through 7 of the state display photograph in Fig. 6-6(a) shows that these instructions have been executed in the proper sequence.

The instruction starting at address 037 is a conditional jump that is a two-word instruction (lines 8 and 9 of the state display). If the chip passed the test (accumulator contains all zeros), the program continues the test routine. If the chip failed the test, the program jumps to an output routine. Examination of line 10 of the state display, address 039, reveals that the chip passed the test. The program then outputs another bit pattern (1111) to the chip under test and reads the input port. This is shown by lines 10 through 13 of the state display. Lines 14 and 15 of the state display are the addresses of the two words of another JCN instruction. Line 16 of the state display is the address 043, showing that the chip failed the last test, causing the program to jump to the output routine.

It is instructive to consider the map display at this time. If a tabular display is not presented in the foregoing procedure, it means that the system did not access the selected address, and the No Trigger light will be on. To find where the system is residing in the program, the operator switches to "map" as exemplified in Fig. 6-7. Using the trigger word switches, he moves the cursor (an illuminated circle) to enclose one of the dots on the screen. Then he switches to Expand and finalizes the cursor position; the No Trigger light will then go out, and switching back to Table A will display the 16 addresses around that point.

When program deviations occur, the reason may be as simple as a program error, or as complicated as a hardware failure on the data

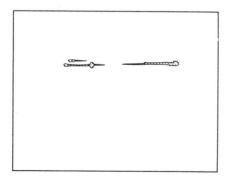

Figure 6-7 A map display shows the entire system activity. (*Courtesy of Hewlett-Packard*)

bus or command lines. Additional input channels now become very desirable. By combining the instruments illustrated in Fig. 6-3, the trigger and display capability can be expanded to 32 bits wide, allowing the 12-bit address, eight-bit data word, and up to 12 other active control signals to be viewed simultaneously. The operator proceeds as follows: (1) Connect data cable between rear panel connectors. (2) Connect trigger bus cable between front panel bus connections. (3) Set the 1600A controls as explained previously, with the following exception: set the display mode to Table A and B. (4) The 1607A controls are set as follows: sample mode, SINGLE; start display, ON; trigger mode with NORM/ARM to NORM, LOCAL/BUS to BUS, and OFF/ WORD to OFF; threshold, logic, and clock as noted previously; all other pushbuttons to their Out positions; qualifiers Q, Q0 to OFF. (5) The operator connects the data and clock inputs for the 1607A as follows: (a) Connect 1607A data inputs 0 through 7 to the demulti- plexed data bus (ROM output) starting with LSB connected to 1607A data input 0; (b) connect 1607A clock input to signal used to clock the 1600A; (c) connect grounds to appropriate points. (6) After a display is on-screen, set the 1607A blanking to display eight columns.

Consider next the display interpretation of the address and data lines. *By displaying both address and data, it is now possible to confirm exact system operation with respect to the test routine.* Looking at line 1 of the state display photograph in Fig. 6-8, observe that the data corresponding to address 030 are F0, the 8-bit code for the CLB instruction. Looking at line 2, one sees that the displayed word agrees with the operation code for the STC instruction given in the program listing. In this manner, subsequent lines of the state display can be examined to show exact program operation. Note that line 14 of the state display corresponds to the first word of the JCN instruction at address 03D. The data in line 15 correspond to the second word (0100 0011) of the JCN instruction, the address to which program control is transferred if the jump condition is true. Examination of line 16 reveals that the program did jump to the specified address.

It is instructive to consider the viewing of the multiplexed data bus. In the preceding examples, the demultiplexed address and data lines have been observed on the ROM address and data lines. However, when a hardware failure occurs, it may be very useful to directly observe activity on the multiplexed microprocessor data bus. In the following example, the reader will observe data being demultiplexed into a 12-bit address for driving a ROM. Then, he will watch the ROM output being multiplexed back onto the data bus. The 1600A and 1607A are set up as follows to obtain the display: (1) Set the 1600A data, qualifier, and clock input with (a) 1600A data inputs 0 through 7 to RD0 through RD7 on the ROM in order; (b) 1600A data inputs 8 through 15 to

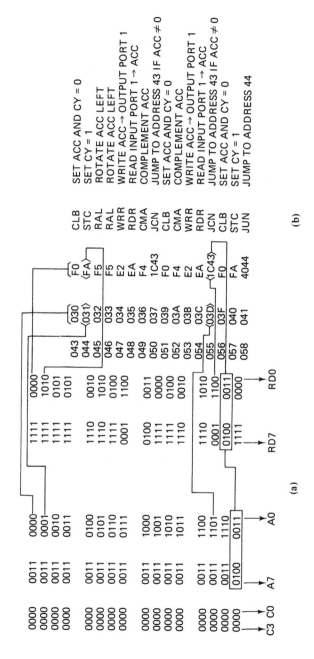

Figure 6-8 System response to test routine on address and data lines. *(Courtesy of Hewlett-Packard)*

145

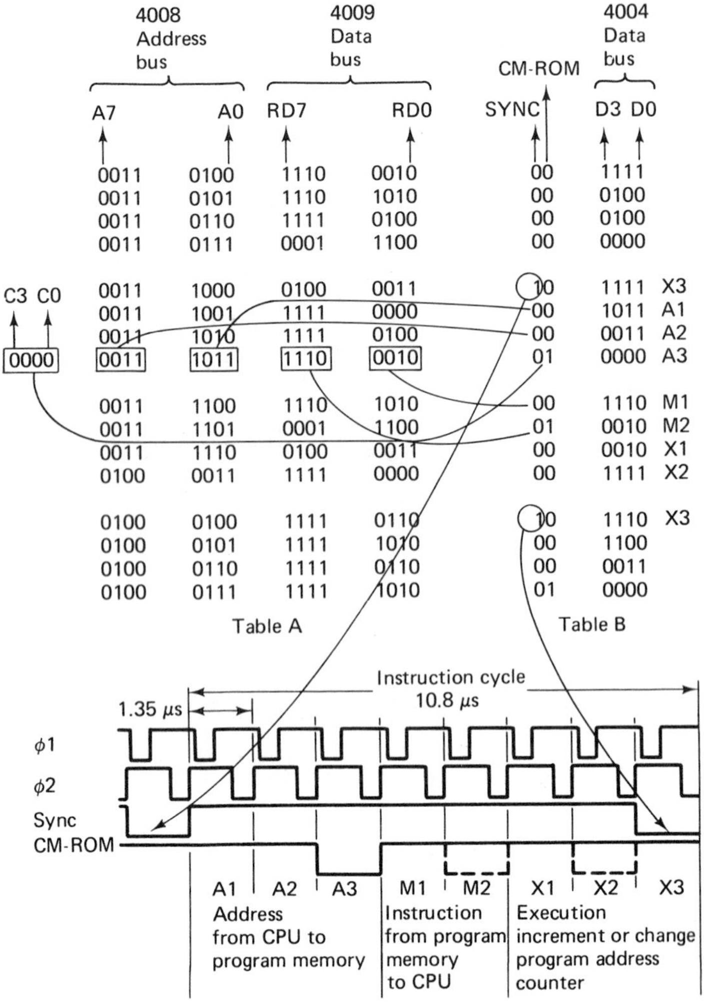
4008 Address bus
A7 A0
4009 Data bus
RD7 RD0
CM-ROM SYNC
4004 Data bus
D3 D0
0011 0100 1110 0010 00 1111
0011 0101 1110 1010 00 0100
0011 0110 1111 0100 00 0100
0011 0111 0001 1100 00 0000
C3 C0
0000
0011 1000 0100 0011 10 1111 X3
0011 1001 1111 0000 00 1011 A1
0011 1010 1111 0100 00 0011 A2
0011 1011 1110 0010 01 0000 A3
0011 1100 1110 1010 00 1110 M1
0011 1101 0001 1100 01 0010 M2
0011 1110 0100 0011 00 0010 X1
0100 0011 1111 0000 00 1111 X2
0100 0100 1111 0110 10 1110 X3
0100 0101 1111 1010 00 1100
0100 0110 1111 0110 00 0011
0100 0111 1111 1010 01 0000
Table A
Table B
Instruction cycle
10.8 μs
1.35 μs
ϕ1
ϕ2
Sync
CM-ROM
A1 A2 A3
Address from CPU to program memory
M1 M2
Instruction from program memory to CPU
X1 X2 X3
Execution increment or change program address counter

Figure 6-9 Comparison of 4004 data bus activity with demultiplexed address and data. (*Courtesy of Hewlett-Packard*)

A0 through A7 in order; (c) 1600A Q0 input to ROM 0 chip select line (1702A, pin 14). Note that by qualifying on $\overline{CS}$ and triggering on A0 through A7, the operator derives a unique trigger that is effectively 12 bits wide with only the eight least significant bits displayed; (d) 1600A clock input, same as previously noted.

(2) Connect the 1607A data and clock inputs to the 4004 microprocessor as follows: (a) 1607A data inputs 0 through 3 to D0 through

D3; (b) 1607A data input 4 to CM-ROM; (c) 1607A data input 5 to SYNC; (d) 1607A clock input to $\phi2$. Set the 1600A controls the same as explained previously, with the following exceptions: display mode, Table A + B; end display, ON; delay, ON with delay set to 8; qualifier, TRIG with Q0 set to LO; column blanking, ccw. Set the 1607A controls as follows: end display, ON; delay, ON with delay set to 8; logic, Neg. Note that the microprocessor data bus uses negative logic; i.e., the most positive voltage is a logic "0," and the most negative voltage is a logic "1." After a display is obtained, the operator adjusts the 1607A column blanking to display six columns in Table B.

Next, consider display interpretation of the multiplexed data bus. The state display in Fig. 6-9 shows a comparison of the demultiplexed address and data buses (Table A) with the multiplexed microprocessor bus (Table B). Compare line 8 of Table A (trigger word) with the multiplexed data in Table B. Examination of the sync line shows that line 6 of Table B corresponds with instruction cycle state Al. Note that the sync and CM-ROM pulses are displayed as one's in the photograph, since negative logic has been selected on the 1607A. Comparison of states A1, A2, and A3 (lines 6, 7, and 8 of the Table B state display) with the trigger word address bits reveals that the demultiplexer has correctly processed the address from the 4004. Similarly, comparison of trigger word data bits RD7 through RD0 with states M1 and M2 (lines 9 and 10 of the Table B display) shows that the multiplexer has correctly processed the ROM data onto the 4004 data bus. Note that the CM-ROM line is true during the M2 state, indicating that the instruction being executed is an I/O instruction.

From the foregoing examples, it may be concluded that *efficient debugging of the 4004 microprocessor system is expedited by two factors: first, the availability of the program listing as produced by the 4004 cross-assembler, and second, the availability of real-time logic state analysis for rapid error detection and correction.*

6-3 EXAMPLE OF EIGHT-BIT MICROPROCESSOR SYSTEM

The Intel 8008 eight-bit microprocessor chip has more computing capability and flexibility than the 4004. It is more suitable for control applications and data handling. The 8008 microprocessor family operates from +5- and −9-V sources. This microprocessor features an eight-bit address and data bus (D_0 through D_7) that, by time multiplexing, allows control information, 14-bit addresses, and eight-bit data bytes to be transmitted between the CPU and the external memory. Pin assignments are depicted in Fig. 6-10. The 14-bit address permits direct addressing of 16,000 words of memory. This microprocessor provides state signals, cycle control signals, and a synchronizing signal to peripheral circuits. These lines are decoded outside the microprocessor to provide the control and timing signals for the microprocessor system. All microprocessor inputs are TTL-compatible and all outputs are low-power TTL-compatible. The microprocessor operates with a 500-kHz clock.

Control lines for the 8008 microprocessor are summarized as follows:

INT: When interrupt (INT) lines are enabled (HIGH), the CPU recognizes the interrupt request at the next instruction fetch cycle.
RDY: HIGH (logic "1") indicates to the CPU that valid memory

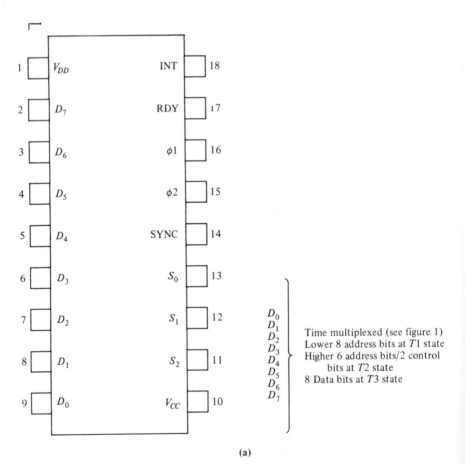

(a)

Figure 6-10   Intel 8008 microprocessor. **(a)** Pin assignments;

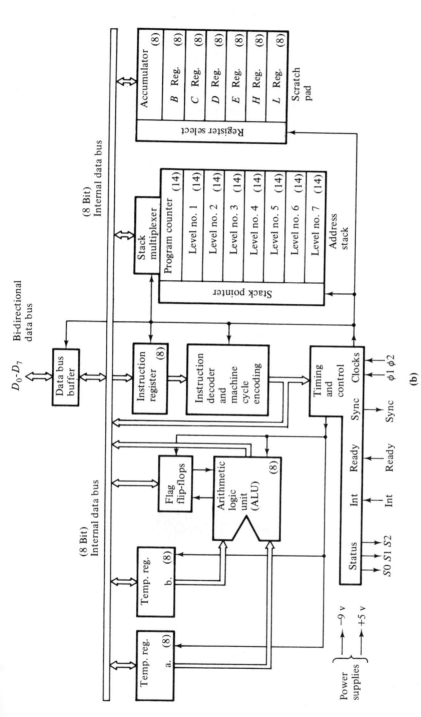

Figure 6-10 Continued **(b)** microprocessor organization.

149

S_0	S_1	S_2	State
0	1	0	$T1$
0	1	1	$T11$
0	0	1	$T2$
0	0	0	Wait
1	0	0	$T3$
1	1	0	Stopped
1	1	1	$T4$
1	0	1	$T5$

Figure 6-11   Tabulation of state control signals.

D_6	D_7	Cycle
0	0	Instruction fetch cycle (PCl)
0	1	Data read (PCR)
1	0	I/O operation (PCC)
1	1	Data write (PCW)

Figure 6-12   Tabulation of cycle control bits.

data are available. LOW (logic "0") indicates to the CPU that valid memory data are not available. SYNC: synchronizing signal indicating the start of each machine state. S_0, S_1, S_2: state control signals. S_0, S_1, and S_2 control use of the data bus and indicate the state of the CPU to the peripheral circuitry. These states are tabulated in Fig. 6-11. D_6, D_7: cycle control bits, designating whether cycle is instruction fetch, data read, data write, or I/O operation, at T2 state. These cycles are tabulated in Fig. 6-12.

Probe Connections

A system that will not "come up" can often be debugged by monitoring the address flow alone. Since the 8008 14-bit address is time-multiplexed, external address latches, such as the Intel 3404 latch, are required in an 8008 system. Connect the analyzer probes to the output side of the eight LSB address latches and to the input side of the six MSB address latches and the cycle control bit latches. The analyzer probe connections shown in Fig. 6-13 provide a display of the activity on the address lines. To set the analyzer controls, the operator turns the power on and proceeds as follows: display mode, Table A; sample mode, REPET. Note that if a program is not looping or cycling through the selected address, select SGL, press RESET, and start the system. *The first time that the system passes through the trigger point, the dis-*

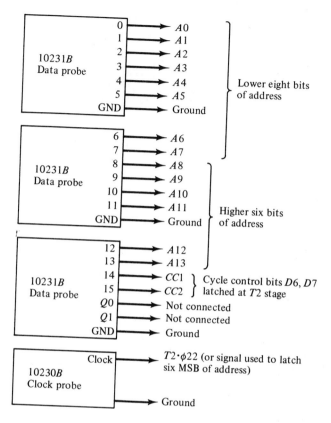

Figure 6-13 Data and clock probe connections. (*Courtesy of Hewlett-Packard*)

play will be generated and stored. The timing diagram for the 8008 microprocessor is shown in Fig. 6-14.

The trigger mode is set with NORM/ARM to NORM, with LOCAL/BUS to LOCAL, and OFF/WORD to WORD. START DSPL is set to ON; CLOCK is set to   . In the system used for this example, the higher six-bit address latches and cycle-code bit latches are clocked on the leading edge of T2•φ22. Note that the clocking requirements of another 8008 system may vary from this example. THLD is set to TTL; the 8008 output lines are low-power TTL-compatible, as are most address latches. If other logic levels are used, set THLD to Variable and adjust the threshold to match the given threshold level. All other pushbuttons are set to their Out positions; display time, ccw; column blanking, ccw; qualifier Q1, Q0 is set to its OFF position;

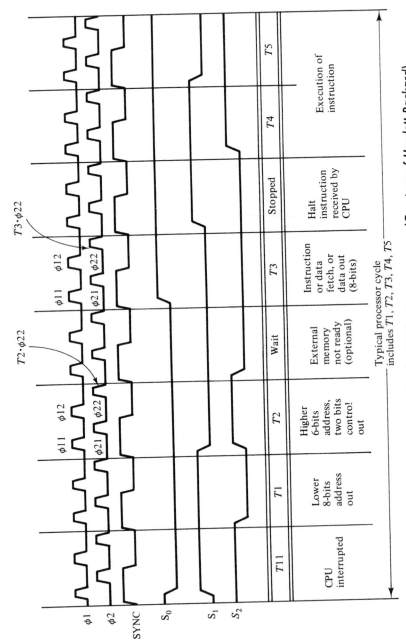

Figure 6-14 Timing diagram for the 8008 microprocessor. (Courtesy of Hewlett-Packard)

trigger word switches are set to match the address on which the designer wishes to trigger.

Consider next the interpretation of the display. In this illustration, system response to a call instruction is considered. The call instruction calls a subroutine to check the keyboard for the presence of a stop command and to check the system status. Proper operation is confirmed by a comparison between real-time state analysis, Fig. 6-15(a), and the 8008 cross assembler listing output, Fig. 6-15(b). The 8008 responds to a call instruction in the following manner: (1) Stores the content of the program in the push-down address stack. (2) Jumps unconditionally to the instruction located in memory location addressed by byte two and byte three of the call instruction. (3) Begins execution of subroutine.

Note the program listing depicted in Fig. 6-15(b). Observe the three-byte call instruction at location 00400. The first byte is the operation code, indicated by 00 in bits 15, 14 columns. The second and third bytes form a double-byte operand (indicated by 01 in bits 15, 14 columns), in this case the address of the first instruction in the subroutine. Proper operation of the Call instruction is confirmed by observing that the address immediately following the third byte of the call instruction, 00402, is 00572. This means that the microprocessor fetched 172 (lower eight bits of subroutine address) from location 00401 and 01 (higher six bits of subroutine address) from location 00402.

$$00\ 000\ 101\ 111\ 010 \leftarrow (000\ 001), (01\ 111\ 010)$$

The MVIH, KYBRD and MVIL, KYBRD instructions (Load Keyboard address in H and L registers) may be confirmed by observing the fourth, fifth, sixth, and seventh lines of the table display photograph. Line 4 is the fetch of the MVIH operation code and line 5 is the fetch of the higher six bits of the keyboard address. Line 6 is the fetch of the MVIL operation code, with line 7 being the fetch of the lower eight bits of the keyboard address. Line 8 is the fetch of operation code for MOV A, M, and line 9 is the fetch of the keyboard character. *In a similar fashion, each instruction in the subroutine may be shown to have been properly executed.*

To view addresses following the last displayed address, simply set the trigger word switches to match the address displayed in line 16. This address becomes the trigger word in line 1, with the next 15 addresses listed in lines 2 through 16. If the operator wishes to retain the original trigger point, an alternate technique is to use digital display and set the thumbwheels to 00015, which provides the same display.

Consider next the selective store function. It may be desirable

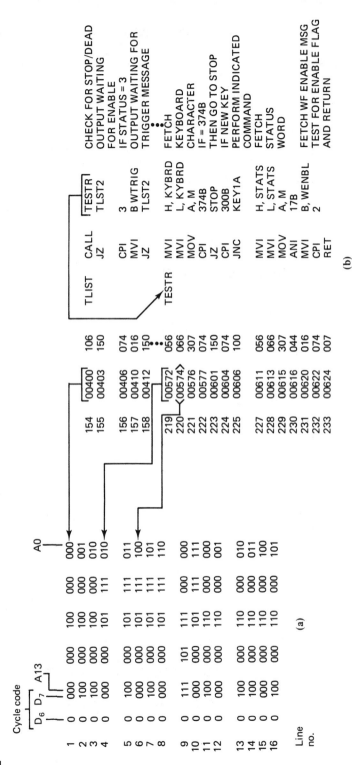

Figure 6-15 System response to CALL instruction. (*Courtesy of Hewlett-Packard*)

to not look at every address, but only those corresponding to instruction fetch cycles. The operator can do this by using the analyzer's display qualifier feature. Looking back at the sample program, Fig. 6-6(b), observe that the subroutine is 14 instructions long with each instruction in the subroutine requiring at least two memory locations. In turn, the operator cannot view the entire subroutine on the 16-word display in Fig. 6-15. By qualifying the display on the two cycle-control bits, it is possible to look at only addresses corresponding to instruction fetch cycles. The operator can do this in the following manner:

1. Connect Q1 and Q0 probes to monitor cycle control bits D_6 and D_7.
2. Set DSPLY/TRIG pushbutton to DSPLY.
3. Set Q1 and Q0 switch to LO.

The state display shown in Fig. 6-16(a) is then obtained. Bits 15 and 14 are both zero for every displayed address, indicating that each displayed address represents an instruction fetch. Comparing the table display with the program listing reveals that line 1 is the address of the call instruction, lines 2 through 15 denote the subroutine, and line 16 is the return to the main program. In turn, the operator has an overview of the entire subroutine.

Next, it is instructive to consider the map. If a tabular display is not presented in the foregoing procedure, it means that the system did not access the selected address, and the No Trigger light will be on. To find where the system is residing in the program, switch to "map" (Fig. 6-17). Using the trigger word switches, move the cursor to encircle one of the dots on screen. Switch to Expand and make the final positioning of the cursor. The No Trigger light will then go out, and switching back to Table A displays the 16 addresses around that point.

When program deviations are found, the reason may be as simple as a program error or as complicated as a hardware failure on the data/control bus, or other command lines. It is helpful in this situation to employ additional input channels by combining the 1600A and 1607A analyzers. This expands the trigger capability to 32 bits wide, allowing the 14-bit addresses, 8 bits of data, and up to ten other active command signals to be viewed simultaneously. Observe the sample program in Fig. 6-18(b). By displaying both address and data, it is now possible to confirm exact system operation with respect to the Call instruction. Looking at line 1 of the state display in Fig. 6-18(a), observe that bits 15 and 14 of the left-hand table are 00, indicating that the eight bits of displayed data represent an operation code. Code 01 000 110 is the call instruction. The second and third byte of a call instruction should be

Figure 6-16 (a)

Line no.	Cycle code D6	D7	A13		A0				Address
1	0	000	000	100	000	000	000		00400
2	0	000	000	101	111	101	111	010	00403
3	0	000	000	101	111	101	111	100	
4	0	000	000	101	111	101	111	110	
5	0	000	000	101	111	101	110	111	00406
6	0	000	000	110	000	110	110	001	00410
7	0	000	000	110	000	110	110	100	00412
8	0	000	000	110	000	110	110	110	
9	0	000	000	110	001	110	110	001	00572
10	0	000	000	110	001	110	110	011	00574
11	0	000	000	110	001	110	110	101	00576
12	0	000	000	110	001	110	100	110	00577
13	0	000	000	110	010	110	110	000	00611
14	0	000	000	110	010	110	110	010	00613
15	0	000	000	110	010	110	010	100	00615
16	0	000	000	100	000	100	000	011	00624

Figure 6-16 (b)

Addr	Oct	Label	Mnemonic	Operand	Comment
00400	106	TLIST	CALL	TESTR	CHECK FOR STOP/DEAD
00403	150		JZ	TLST2	OUTPUT WAITING FOR EANBLE
00406	074		CPI	3	IF STATUS = 3
00410	016		MVI	B, WTRIG	OUTPUT WAITING FOR
00412	150		JZ	TLST2	TRIGGER MESSAGE
	•••				•••
00572	056	TESTR	MVI	H, KYBRD	FETCH
00574	066		MVI	L, KYBRD	KEYBOARD
00576	307		MOV	A, M	CHARACTER
00577	074		CPI	374B	IF = 374B
00601	150		JZ	STOP	THEN GO TO STOP
00604	074		CPI	300B	IF NEW KEY
00606	100		JNC	KEY1A	PERFORM INDICATED COMMAND
00611	056		MVI	H, STATS	FETCH
00613	066		MVI	L, STATS	STATUS
00615	307		MOV	A, M	WORD
00616	044		ANI	17B	FETCH W.F. ENABLE MSC
00620	016		MVI	B, WENBL	TEST FOR EANBLE FLAG
00622	074		CPI	2	
00624	007		RET		AND RETURN

Figure 6-16 Qualified display showing the ability to selectively display only desired addressed data. (Courtesy of Hewlett-Packard)

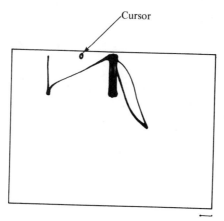

Figure 6-17 Map display shows entire system activity. (*Courtesy of Hewlett-Packard*)

the lower and upper address bits, respectively, of the subroutine being called. Examination of the address in the fourth line reveals that, indeed, the data bytes of lines 3 and 2 (00 000 001 and 01 111 010) have been combined to form the subroutine address (00 000 101 111 010). In a similar manner, each line of the display can be examined to reveal exact program operation.

6-4 INTEL 4040 MICROPROCESSOR SYSTEMS

The 4040 microprocessor, which is the core of the 4040 microcomputer family, is fabricated with P-channel silicon-gate MOS technology and operates from +5- and −10-V power supplies. The 4040 features a four-bit parallel CPU with 60 instructions. It can directly address 4000 eight-bit instruction words of program memory or 8k with bank switching, and 5120 bits of data storage RAM. Up to 16 four-bit input ports and 16 four-bit output ports can also be addressed directly. Twenty-four randomly accessible index registers are provided internal to the microprocessor for temporary data storage. This microprocessor operates at clock rates up to approximately 750 kHz. Pin assignments for the microprocessor package are shown in Fig. 6-19.

Control lines are summarized as follows:

STPA, signal acknowledges that the processor has entered its Stop mode; STP, a logic "1" on this input causes the processor

Figure 6-18 — assembly listing (System response to CALL on address, data, and control lines)

Address	Line	Label	Mnemonic	Operand	Comment
00400	154	TLIST	CALL	TESTR	CHECK FOR STOP/DEAD
00403	155		JZ	TLST2	OUTPUT WAITING FOR ENABLE
00406	156		CPI	3	IF STATUS = 3
00410	157		MVI	B, WTRIG	OUTPUT WAITING FOR
00412	158		JZ	TLST2	TRIGGER MESSAGE
00572	219	TESTR	MVI	H, KYBRD	FETCH
00574	220		MVI	L, KYBRD	KEYBOARD
00576	221		MOV	A, M	CHARACTER
00577	222		CPI	374B	IF = 374B
00601	223		JZ	STOP	THEN GO TO STOP
00604	224		CPI	300B	IF NEW KEY
00606	225		JNC	KEY1A	PERFORM INDICATED COMMAND
00611	227		MVI	H, STATS	FETCH
00613	228		MVI	L, STATS	STATUS
00615	229		MOV	A, M	WORD
00616	230		ANI	17B	
00620	231		MVI	B, WENBL	FETCH W.F. ENABLE MSG
00622	232		CPI	2	TEST FOR ENABLE FLAG
00624	233		RET		AND RETURN

Figure 6-18 System response to CALL on address, data, and control lines. (Courtesy of Hewlett-Packard)

to enter its Stop mode; INT, a logic "1" on this input causes the processor to enter its interrupt mode; INTA, signal acknowledges receipt of an interrupt command and prevents additional interrupts from entering the processor. Signal remains active until cleared by the BBS instruction. $\phi 1$, $\phi 2$ are nonoverlapping clock signals that determine the microprocessor timing. RESET, a "1"

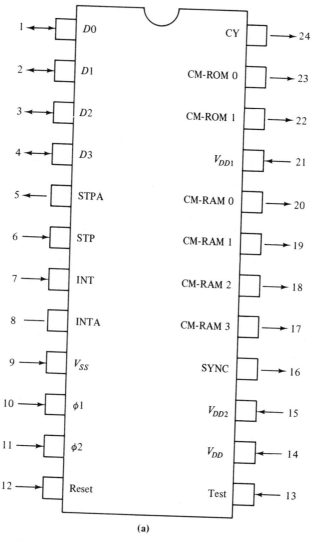

(a)

Figure 6-19 Intel 4040 microprocessor. (a) Pin assignments;

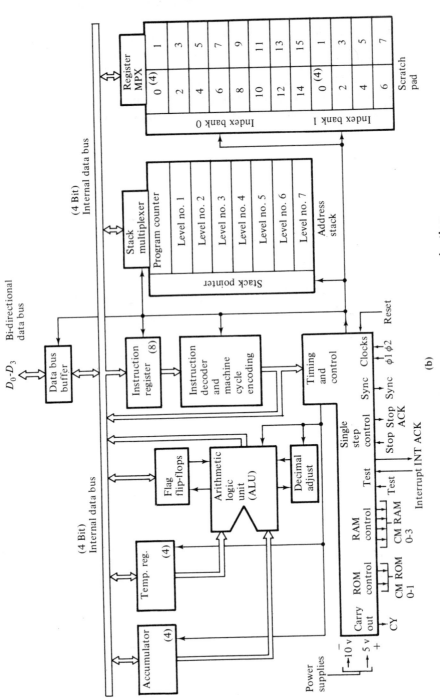

Figure 6-19 Continued   **(b)** microprocessor organization.

level applied to RESET clears all flag and status flip-flops and forces the program counter to 0. RESET must be applied for 96 clock cycles (12 machine cycles) to completely clear all address and index registers. Test, input. The logic state of TEST can be examined with JCN instruction. SYNC, synchronization signal indicating beginning of instruction cycle to ROM and RAM chips. CM-RAM 0 through CM-RAM 3, lines function as bank select signals for the RAM chips in the system. CM-ROM 0 and CM-ROM 1, bank selection signals for program ROM chips in the system. CY, the state of the carry flip-flop is present on this output and is updated each X1 time.

Probe connections are depicted in Fig. 6-20. A system that will not "come up" can often be debugged by monitoring address flow alone. The 4040 CPU chip has a four-bit data bus, on which the 12-bit address is multiplexed during A1, A2, and A3 states of the 4040 machine cycle. To observe the demultiplexed 12-bit address on a 1600A, the 4040 system must use the 4008/4009 Standard Memory and I/O Interface Sets, 4289 Standard Memory Interfaces, or similar logic circuits that provide a demultiplexed address bus. If the designer's system uses memory chips that internally decode the multiplexed address, such as the 4001 ROM, the operator can monitor the microprocessor data bus as detailed subsequently.

Note that although the 4040 microprocessor does not provide a unique clock for the logic state analyzer at the proper time (end of A3 state) in the instruction cycle, the CM-ROM line is always true at A3 and can be used as a clock signal. However, CM-ROM also occurs at states M1, M2, and X2 during the execution of some instructions. This would result in invalid data's being displayed by the analyzer. By employing the circuit depicted in Fig. 6-21, the operator can ensure a correct state display. Use CM-ROM 0 if monitoring the program stored in ROM bank 0, or CM-ROM 1 if monitoring the program stored in ROM bank 1.

Consider the display interpretation. In this illustration, an output routine is examined. Proper operation is confirmed by a comparison between real-time state analysis, Fig. 6-22(a), and the 4040 cross assembler program listing, Fig. 6-22(b). The output routine performs the following functions: (1) Sets up a bit pattern in the accumulator. (2) Outputs the accumulator contents to an I/O port for control of status lights. (3) Reads status of start switch connected to input port. (4) If switch position is true (CY = 1), clears status lights and jumps to start routine; if switch position is false (CY = 0), loops.

Refer to Fig. 6-22(a). Line 1 displays the address 0000 0100

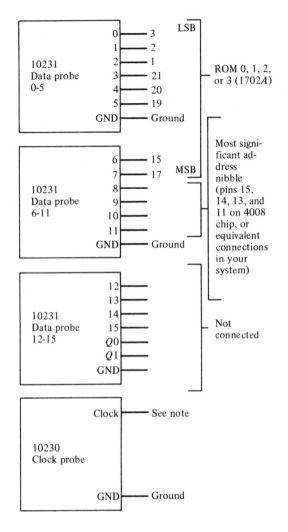

Figure 6-20 Data probe connections. (*Courtesy of Hewlett-Packard*)

0000 (040). This corresponds to the address of the STC instruction included in the program listing, Fig. 6-22(b). Lines 2 and 3 of the state display correspond to the JUN instruction at address 041. JUN is a two-word instruction, thus occupying two address locations. Examination of line 4 of the state display (0000 0100 0100) shows that the JUN instruction was properly executed; that is, the program jumped to address 044. In a similar fashion, each instruction can be shown to have been executed in the proper sequence. The last address shown in the

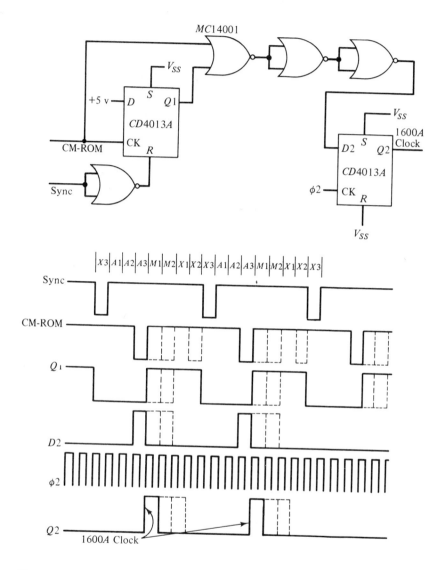

Figure 6-21 Circuit for deriving a clock for the 1600A from 4040 Sync, CM-ROM, and $\phi 2$ signals. (*Courtesy of Hewlett-Packard*)

state display is address 050. To view subsequent addresses, the operator sets the trigger word switches to match address 050. This address would then become the trigger word in line 1, with the next 15 addresses listed in lines 2 through 16. *If the operator wishes to retain the original trigger*

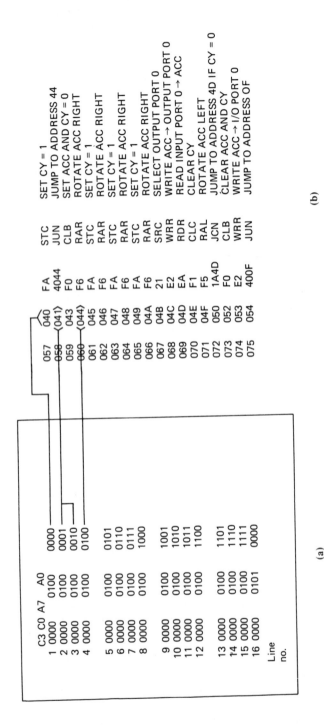

C3	C0	A7	A0
1	0000	0100	0100	0000
2	0000	0100	0100	0001
3	0000	0100	0100	0010
4	0000	0100	0100	0100
5	0000	0100	0100	0101
6	0000	0100	0100	0110
7	0000	0100	0100	0111
8	0000	0100	0100	1000
9	0000	0100	0100	1001
10	0000	0100	0100	1010
11	0000	0100	0100	1011
12	0000	0100	0100	1100
13	0000	0100	0100	1101
14	0000	0100	0100	1110
15	0000	0100	0100	1111
16	0000	0100	0101	0000

Line no.

(a)

057	040	FA	STC	SET CY = 1
058	(041)	4044	JUN	JUMP TO ADDRESS 44
059	043	F0	CLB	SET ACC AND CY = 0
060	(044)	F6	RAR	ROTATE ACC RIGHT
061	045	FA	STC	SET CY = 1
062	046	F6	RAR	ROTATE ACC RIGHT
063	047	FA	STC	SET CY = 1
064	048	F6	RAR	ROTATE ACC RIGHT
065	049	FA	STC	SET CY = 1
066	04A	F6	RAR	ROTATE ACC RIGHT
067	04B	21	SRC	SELECT OUTPUT PORT 0
068	04C	E2	WRR	WRITE ACC → OUTPUT PORT 0
069	04D	EA	RDR	READ INPUT PORT 0 → ACC
070	04E	F1	CLC	CLEAR CY
071	04F	F5	RAL	ROTATE ACC LEFT
072	050	1A4D	JCN	JUMP TO ADDRESS 4D IF CY = 0
073	052	F0	CLB	CLEAR ACC AND CY
074	053	E2	WRR	WRITE ACC → I/O PORT 0
075	054	400F	JUN	JUMP TO ADDRESS OF

(b)

Figure 6-22 System response to output routine. (*Courtesy of Hewlett-Packard*)

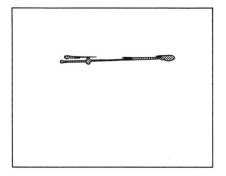

Figure 6-23 Map display reveals the entire system activity. (*Courtesy of Hewlett-Packard*)

point, an alternate technique is to employ digital delay and to set the thumbwheels to 00015, which would provide the same display.

Next, observe the map display exemplified in Fig. 6-23. If a tabular display is not obtained in the foregoing procedure, it means that the system did not access the selected address, and the No Trigger light will be on. To determine where the system is residing in the program, the operator switches to "map," as shown in Fig. 6-23. Using the trigger word switches, he moves the cursor to encircle one of the dots on the screen. Then, he switches to Expand and finalizes the cursor position. In turn, the No Trigger light will go out, and switching back to Table A will display the 16 addresses around that point. *If a deviation is indicated, the fault may be in program preparation, or it may be a hardware failure.* To track down the difficulty, it is very helpful to have additional input channels. This increased capability is obtained by combining the 1600A and 1607A analyzers.

With reference to Fig. 6-24, it is instructive to observe the display interpretation of address, data, and control lines. By displaying address, data and control lines, the designer is enabled to confirm exact system operation with respect to the output routine. Consider the program listing exemplified in Fig. 6-24(b). The first address, 040, contains an STC instruction. The second address, 041, is the address of the first word of a two-word JUN instruction. These two instructions are shown in lines 1 through 3 of the state display in Fig. 6-24(a). Line 1 shows the address (0000 0100 0000) and instruction code (1111 1010) of the STC instruction. Lines 2 and 3 show the addresses of the two words that make up the JUN instruction. The first byte (0100) of the JUN instruction is the operation code and the remaining three bytes (0000

0100 0100) constitute the address to which program control is transferred.

Examination of line 4 of the state display shows that program was actually transferred to the specified address that contains an RAR instruction (instruction code 1111 0110). Line 5 of the state display corresponds to the STC (Set Carry) instruction at address 045. Proper execution of the STC instruction is confirmed by observing that the carry bit (CY column) in line 6 of the state display is a one. The one

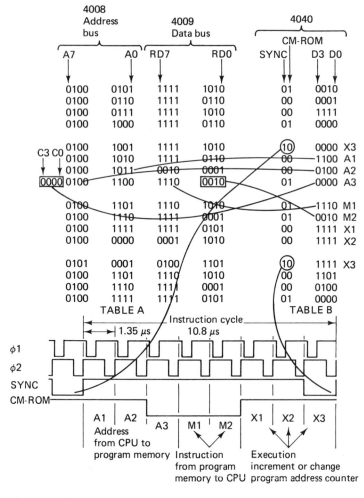

Figure 6-24   System response to output routine on address, data, and control lines. (*Courtesy of Hewlett-Packard*)

Figure 6-24 Continued — 4040 data bus activity with demultiplexed address and data

(a)

Line no.	C3 C0	A7	A0	CY	RD7	RD0	Addr	Prog. Addr.	Hex
1	0000	0100	0000	1	1111	1010	057		FA
2	0000	0100	0001	1	0100	0000	058	⟨040⟩ ⟨041⟩	4044
3	0000	0100	0010	0	0100	0100	059	043	F0
4	0000	0100	0100	0	1111	0110	060	044	F6
5	0000	0100	0101	0	1111	1010	061	045	FA
6	0000	0100	0110	1	1111	0110	062	046	F6
7	0000	0100	0111	0	1111	1010	063	047	FA
8	0000	0100	1000	1	1111	0110	064	048	F6
9	0000	0100	1001	0	1111	1010	065	049	FA
10	0000	0100	1010	1	1111	0110	066	04A	F6
11	0000	0100	1011	0	0010	0001	067	04B	21
12	0000	0100	1100	1	1110	0010	068	04C	E2
13	0000	0100	1101	1	1110	1010	069	04D	EA
14	0000	0100	1110	1	1111	0001	070	04E	F1
15	0000	0100	1111	1	1111	0101	071	04F	F5
16	0000	0100	0000	1	0001	1010	072	050	1A4D
							073	052	F0
							074	053	E2
							075	054	400F

(b)

Hex	Mnemonic	Description
FA	STC	SET CY = 1
4044	JUN	JUMP TO ADDRESS 44
F0	CLB	SET ACC AND CY = 0
F6	RAR	ROTATE ACC RIGHT
FA	STC	SET CY = 1
F6	RAR	ROTATE ACC RIGHT
FA	STC	SET CY = 1
F6	RAR	ROTATE ACC RIGHT
FA	STC	SET CY = 1
F6	RAR	ROTATE ACC RIGHT
21	SRC	SELECT OUTPUT PORT 0
E2	WRR	WRITE ACC → OUTPUT PORT 0
EA	RDR	READ INPUT PORT 0 → ACC
F1	CLC	CLEAR CY
F5	RAL	ROTATE ACC LEFT
1A4D	JCN	JUMP TO ADDRESS 4D IF CY = 0
F0	CLB	CLEAR ACC AND CY
E2	WRR	WRITE ACC → I/O PORT 0
400F	JUN	JUMP TO ADDRESS 0F

(Courtesy of Hewlett-Packard)

shows up in the CY column one cycle after the STC instruction because the 1600A is clocked during A3 state; in other words, it is clocked before the instruction is executed. In turn, the 1600A does not "see" the results of the instruction execution until the next A3 state. Each line of the display can be examined in a similar manner to reveal exact program operation.

Next, it is helpful to observe the display interpretation of the multiplexed data bus. The state display photograph exemplified in Fig. 6-24 shows a comparison of the demultiplexed address and data buses (Table A) with the multiplexed microprocessor bus (Table B). Compare line 8 of Table A (trigger word) with the multiplexed data in Table B. Examination of the sync line shows that line 6 of Table B corresponds with instruction cycle state A1. Note that the sync and CM-ROM pulses are displayed as ones in the photograph, since negative logic is being employed by the analyzer. Comparison of states A1, A2, and A3 (lines 6, 7, and 8 of the Table B state display) with the trigger word address bits reveals that the interface circuit has correctly demultiplexed the address from the 4040. Similarly, comparison of trigger word data bits RD7 through RD0 with states M1 and M2 (lines 9 and 10 of the Table B display) shows that the interface circuit has correctly demultiplexed the ROM data onto the 4040 data bus. Note that the CM-ROM line is true during M1 and M2 states, indicating that the instruction word being executed is the first word of an I/O instruction.

6-5 INTEL 8080 MICOPROCESSOR SYSTEMS

The 8080 microprocessor is the core of the 8080 family, and is constructed with NMOS technology. It operates from $+12$-, $+5$-, and -5-V sources. Features of the 8080 include an eight-bit, bidirectional, three-state data bus and a separate 16-bit, three-state address bus. The 16-bit address bus permits direct addressing of 65,000 words of memory. Six timing and control outputs are available, whereas four control units and two clock signals are required. All buses are TTL compatible; the microprocessor operates with a 2-MHz clock. Pin assignments are shown in Fig. 6-25.

Pin functions for the 8080 microprocessor are as follows: A_{15}–A_0, address to memory or I/O device number for up to 256 input and 256 output devices; A_0 is LSB. D_7–D_0, bidirectional communication between memory, CPU, and I/O devices. Sync, signal to indicate beginning of each machine cycle. DBIN, data bus in signal indicates to external circuits that data bus is in input mode (enables gating data from I/O or memory onto data bus). READY, valid memory or input data

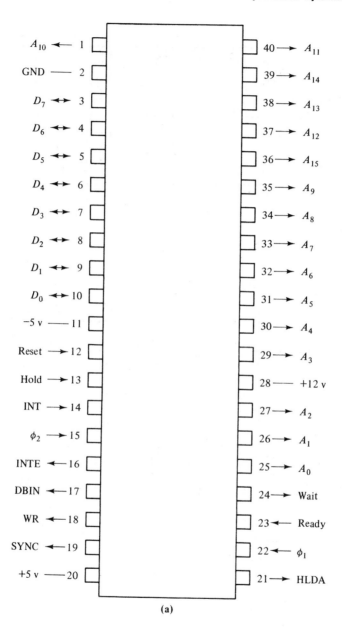

(a)

Figure 6-25 Intel 8080 microprocessor. **(a)** Pin assignments;

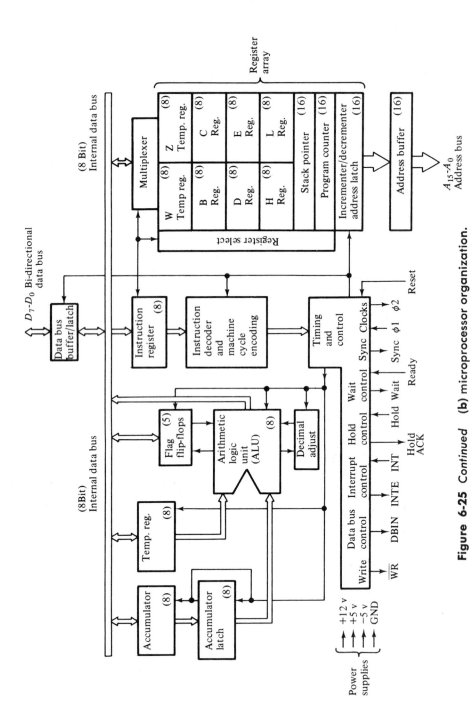

Figure 6-25 Continued **(b)** microprocessor organization.

available on data bus; used to synchronize the microprocessor with slower memory or I/O devices. WAIT acknowledges that the microprocessor is in a wait state. $\overline{WR}$, the WRITE signal used for memory write or I/O output control. HOLD requests microprocessor to enter HOLD state. HLDA, the HOLD ACKNOWLEDGE signal, responds to HOLD and indicates that address and data buses will go to high-impedance state. INTE, Interrupt Enable signal, indicates content of internal interrupt enable flip-flop; inhibits interrupt when flip-flop is reset. INT, interrupt request, is recognized at end of current instruction or while halted. RESET, while activated, the program counter is reset; program will start at 0 in memory; INTE and HLDA are reset. 01 and 02, external clocks; non-TTL compatible.

Data analyzer probe connections are shown in Fig. 6-26. A microprocessor system that will not "come up" can often be debugged by monitoring the address flow alone. Display interpretation is as follows: The CALL instruction initiates a subroutine to check the keyboard for the presence of a stop command and to check system status. Proper operation is confirmed by a comparison of real-time state analysis [Fig. 6-27(a)], and the microprocessor cross assembler listing output [Fig. 6-27(b)]. The 8080 responds to a CALL instruction in the following manner:

1. *Stores content of the program counter in the push-down address stack.*
2. *Jumps unconditionally to the instruction in memory location addressed by byte two and byte three of the CALL instruction.*
3. *Begins execution of the subroutine.*

Examine the program listing in Fig. 6-27(b). Line 1 displays the address of the CALL instruction, 00F8. Therefore, during the next four machine cycles the following actions occur: the high-order eight bits of the next instruction address are stored in the push-down stack; the low-order eight bits of the next instruction address are stored in the push-down stack; the stack pointer is decremented by 2; control is transferred to the first address of the subroutine. Proper operation of the CALL instruction is confirmed by observing that the address on line 6 of the display is the address of the first subroutine instruction. Again, there are multiple cycle operations between the first and second subroutine instructions. *In similar fashion, address lines may be compared with the microprocessor instruction set operation to verify proper subroutine operation.*

To view addresses following the last displayed address (view the next "page"), the operator sets the trigger word switches to match the

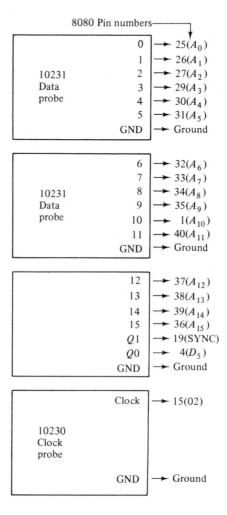

Figure 6-26   Probe connections for data analyzer. (*Courtesy of Hewlett-Packard*)

address displayed in line 16. This address then becomes the trigger word in line 1 of the display, with the next 15 addresses displayed on lines 2 through 16. Or, if the operator wishes to retain the original address word on the trigger word switches, an alternate method is to set the digital delay to 00015 to obtain the desired display. It may be desirable not to look at every address, but only those corresponding to instruction fetch cycles. The operator can do this by using the logic state analyzer "qualifier" feature. With reference to the sample program in Fig. 6-27(b), it is apparent that the subroutine is 10 instructions in length, with most instructions requiring at least two memory locations. Because of this situation, the operator cannot view the entire subroutine on the 16-word display in Fig. 6-27.

By qualifying the display on instruction fetch cycles, it is possible to look at addresses corresponding only to instruction fetch operations. This is accomplished as follows: (1) Connect Q1 line from the 10231 Data Probe to pin 4 (D_5) of the microprocessor. (2) Set Q1 switch to HI. D_5 line of the microprocessor goes HI at the beginning of each instruction fetch cycle, and the operator then obtains the display shown in Fig. 6-28(a). Comparing the table display with the program listing reveals that line 1 is the address for the CALL instruction, lines 2 through 11 are the subroutine, and line 12 is the return to the main program. *This provides the operator with an overview of the entire subroutine.*

If a tabular display is not obtained in the foregoing procedure, it is concluded that the system did not access the selected address, and the No Trigger light will be on. To find where the system is residing in the program, the designer can switch to MAP display (Fig. 6-29). Using the trigger word switches, he moves the cursor to encircle one of the dots on the screen. Then he switches to EXPAND and finalizes the position of the cursor. The No Trigger light will go out, and depressing Table A will provide a display of the 16 addresses around that point. To track down a program deviation, the 1600A and 1607A can be combined to make the display and trigger capability two bits wide, allowing the eight-bit address, eight bits of data, and up to eight other active command signals to be viewed simultaneously.

It is instructive to observe the CALL program instruction in Fig. 6-30(b), and the display in Figure 6-30(a). By displaying both address and data, it is possible to confirm system operation with respect to the CALL instruction. The single bit on the right table display indicates that all Read instructions are being displayed. Looking at line 1 of the left table display, observe that the address, 00F8, corresponds to the program address for the CALL instruction. Also on line 1, the data display corresponds to the operating code, CD, for the subroutine being called. The address on line 2 shows the data bus reading the least significant eight bits of the subroutine address, whereas line 3 shows the data bus reading the most significant bits of the subroutine address. Since the control line is qualified to show only read operations, the designer does not see the program counter being written into the push-down stack memory. Line 4 confirms that the subroutine address is called, and the operating code for a keyboard fetch is displayed on the data lines. *In a similar manner, the designer can examine each line of the display to check the exact operation of the system.*

(a)

Line no.	A$_{15}$–A$_{12}$	A$_{11}$–A$_{8}$	A$_{7}$–A$_{4}$	A$_{3}$–A$_{0}$	
1	0000	0000	1111	1000	} 3 BYTE CALL
2	0000	0000	1111	1001	} STORE ADDR ON STACK
3	0000	0000	1111	1010	
4	0011	0111	1111	1101	
5	0011	0111	1111	1100	} RETRIEVE ADDR
6	0000	0001	0110	1101	} OF SUBROUTINE
7	0000	0001	0110	1110	
8	0000	0001	0110	1111	
9	1111	1011	1100	0000	
10	0000	0001	0111	0000	
11	0000	0001	0111	0001	
12	0000	0001	0111	0010	
13	0000	0001	0111	0011	
14	0000	0001	0111	0100	
15	0000	0001	0111	0101	
16	0000	0001	0111	0110	

(b)

Dec	Addr	Code	Label	Mnemonic	Operand	Comment
151	00F8	CD	TLIST	CALL	TESTR	CHECK FOR STOP/DEAD
152	00FB	CA		JZ	TLST2	OUTPUT WAITING FOR ENABLE
153	00FE	FE		CPI	3	IF STATUS = 3
154	0100	0E		MVI	WTRIG	OUTPUT WAITING FOR
155	0102	CA		JZ	TLST2	TRIGGER MESSAGE
				••••		••••
206	016D	3A	TESTR	LDA	KYBRD	FETCH KEYBOARD CHARACTER
207	0170	FE		CPI	FC	IF = 374B
208	0172	D2		JZ	STOP	THEN GO TO STOP
209	0175	FE		CPI	C0	IF NEW KEY
210	0177	D2		J	KEY1A	PERFORM INDICATED COMMAND
212	017A	3A		LDA	STATS	FETCH STATUS WORD
213	017D	E6		ANI	OF	
214	017F	0E		MVI	CWENBL	FETCH W.F. ENABLE MESSAGE
215	0181	FE		CPI	2	TEST FOR ENABLE FLAG
216	0183	C9		RET		AND RETURN

Figure 6-27 System response to CALL instruction. (*Courtesy of Hewlett-Packard*)

174

(a)

ADDRESS (A15 → ... → A0)

Line no.	A15–A12	A11–A8	A7–A4	A3–A0
1	0000	0000	1111	1000
2	0000	0001	0110	1101
3	0000	0001	0111	0000
4	0000	0001	0111	0010
5	0000	0001	0111	0101
6	0000	0001	0111	0111
7	0000	0001	0111	1010
8	0000	0001	0111	1101
9	0000	0001	0111	1111
10	0000	0001	1000	0001
11	0000	0001	1000	0011
12	0000	0000	1111	1011
13	0000	0001	0000	1010
14	0000	0111	0001	0011
15	0000	0111	0001	0101
16	0000	0111	0001	1000

(b)

Line	Address	Hex	Label	Mnemonic	Operand	Comment
151	00F8	CD	TLIST	CALL	TESTR	CHECK FOR STOP/DEAD
152	00FB	CA		JZ	TLST2	OUTPUT WAITING FOR ENABLE
153	00FE	FE		CPI	3	IF STATUS = 3
154	0100	0E		MVI	WTRIG	OUTPUT WAITING FOR TRIGGER MESSAGE
155	0102	CA		JZ	TLST2	
206	016D	3A	TESTR	LDA	KYBRD	FETCH KEYBOARD CHARACTER
207	0170	FE		CPI	FC	IF = 374B
208	0172	D2		JZ	STOP	THEN GO TO STOP
209	0175	FE		CPI	CO	IF NEW KEY
210	0177	D2		J	KEY1A	PERFORM INDICATED COMMAND
212	017A	3A		LDA	STATS	FETCH STATUS WORD
213	017D	E6		ANI	OF	
214	017F	0E		MVI	CWENDL	FETCH W.F. ENABLE MESSAGE
215	0181	FE		CPI	2	TEST FOR ENABLE FLAG
216	0183	C9		RET		AND RETURN

Figure 6-28 Qualified display showing the ability to display selectively only desired address data. (*Courtesy of Hewlett-Packard*)

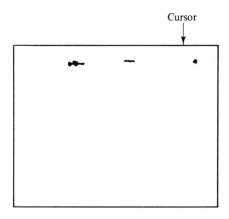

Figure 6-29   Entire system activity is shown by the map display. (*Courtesy of Hewlett-Packard*)

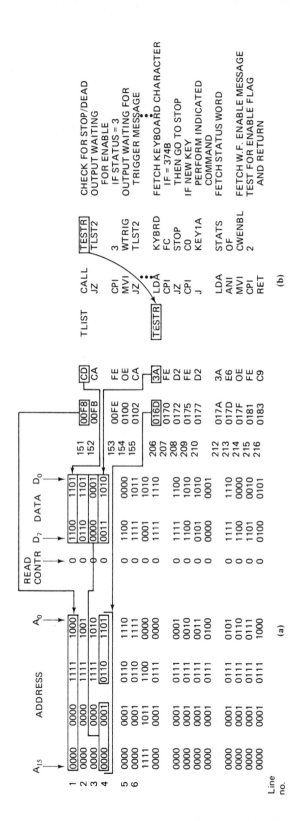

Figure 6-30 System response to CALL on address, data, and control line. (Courtesy of Hewlett-Packard)

7

High-fidelity Systems

7-1 GENERAL CONSIDERATIONS

Although no industry standards have been established, it is generally agreed that high-fidelity reproduction involves a frequency response that is uniform within ±1 dB from 20 Hz to at least 20 kHz, with a harmonic distortion of less than 1 percent. Component systems such as that depicted in Fig. 7-1 are very popular. Thus, a chosen set of stereo-quadraphonic speakers may be utilized with a preferred type of two four-channel amplifiers, plus a selected type of record player (turntable), a chosen design of AM-FM tuner, a selected reel-to-reel tape deck and/or an eight-track deck, or a preferred type of cassette deck. Also, hi-fi stereo-quad systems are also designed in unitized form and housed in elaborate furniture cabinets. A unitized system is termed a *console,* and it contains at least two speakers.

Another type of stereo-quad design, termed the *compact,* has separate speakers, with a record turntable and stereo amplifier on the same base. The main unit in a compact may include an AM and/or FM tuner, plus a turntable. Another design of compact has a record changer mounted on top of the main unit, with a clear plastic cover. A compact is sometimes called a *modular* hi-fi system. A hi-fi speaker enclosure is usually designed with several speaker units. The largest speaker in a group is termed a *woofer,* and it is employed for reproduction of the low bass tones. On the other hand, the smallest speaker, termed a *tweeter,* reproduces the high treble tones. An intermediate size of speaker, often called a *squawker,* or midrange speaker, is gener-

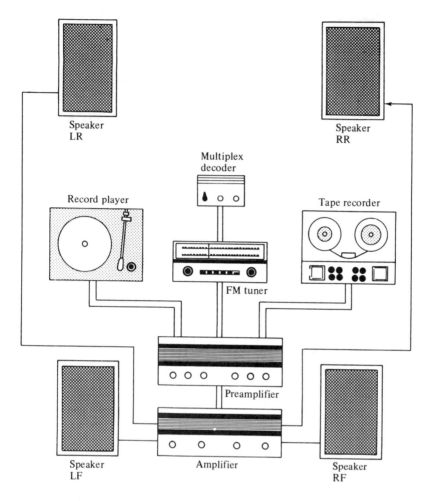

Figure 7-1 Typical high-fidelity component system.

ally utilized to reproduce the middle range of tones between low bass and high treble tones.

Some designs of enclosures contain a pair of midrange speakers, one of which is larger than the other. As a general rule, the size of a speaker is proportional to the amount of audio power that it can accommodate. An enclosure with a woofer, a tweeter, and three midrange speakers is depicted in Fig. 7-2. Since the bass tones in a musical passage are almost always required to carry the major proportion of the total audio power, the woofer is comparatively large. Speakers and

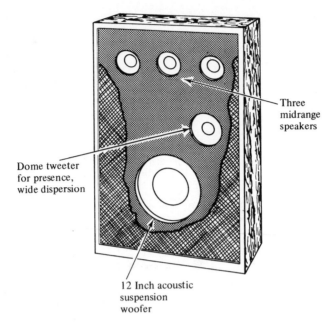

Three midrange speakers

Dome tweeter for presence, wide dispersion

12 Inch acoustic suspension woofer

Figure 7-2 Enclosure with a woofer, a tweeter, and three mid-range speakers.

their associated electrical networks in an enclosure are called a *speaker system*. Network design is discussed in greater detail at a later point in this chapter.

Stereo amplifiers and compact units often provide jacks for plugging in a pair of stereo headphones. Some hi-fi connoisseurs prefer the acoustics of headphones; others utilize them for privacy. Hi-fi amplifiers are designed and rated for very low distortion, for frequency response, for maximum power output, and often for music-power output. Some amplifiers are rated for square-wave response. Various input facilities are provided. For example, an appropriate *input jack* is customarily provided for an AM-FM tuner, for a reel-to-reel tape deck, and for a cassette player. A hi-fi amplifier is also designed with various features, such as tone controls, loudness-type volume control, terminals for additional speakers, stereo balance control, and various input facilities. Hi-fi enthusiasts who make their own tape recordings require an amplifier that provides an appropriate stereo signal for a particular tape recorder.

A *receiver* basically consists of a tuner and an amplifier. Some component systems are designed with a separate tuner and a separate stereo amplifier. All stereo tuners include a *multiplex decoder* section to reconstitute the two stereo signals from the incoming encoded FM signal. A *tape recorder* provides both recording and playback facilities, whereas a *tape player* lacks recording facilities. Note that a *tape deck* is not designed with a built-in amplifier, and is utilized with an external amplifier and speaker system. A tape deck may or may not provide recording facilities. Monophonic recording is accomplished with a single microphone (or audio signal source); stereo recording requires a pair of microphones (or a stereo signal source); quadraphonic recording requires four microphones (or a quadraphonic signal source).

Hi-fi enthusiasts tend to prefer *reel-to-reel* machines over cartridge or cassette-type machines. Eight-track *cartridge* tape players, however, are popular because they are compact and are comparatively simple to operate. The majority of eight-track cartridge tape machines are designed as *player decks*. That is, a player deck lacks recording facilities. All eight-track tape players provide stereo reproduction, and some provide quadraphonic reproduction; many qualify as high-fidelity units.

7-2 BASIC AUDIO AMPLIFIER DESIGN

An audio amplifier is part of a system, and is generally characterized as a preamplifier, driver, or output amplifier, as seen in Fig. 7-3. A typical audio-amplifier channel in a high-fidelity system has a maximum usable gain (MUG) of more than 5000 times, as depicted in Fig. 7-4. The system harmonic distortion is customarily less than one percent at a frequency of 1 kHz. *Power bandwidth* is defined as the frequency range between an upper limit and a lower limit, at a power level 3 dB below maximum rated power output, where the harmonic distortion starts to exceed the value that occurs at 1 kHz and at maximum rated power output. The *music-power* rating of an amplifier is defined as the peak power that can be delivered to the speakers for a very short period of time, with no more harmonic distortion than at the maximum rated sine-wave rms output. That is, a music-power rating denotes the ability of an amplifier to process sudden peak musical waveforms without objectionable distortion. The peak duration in this mode of operation is limited by the capability of the filter capacitors in the power supply to sustain the peak current demand.

Most hi-fi amplifiers can be classified into preamplifier and power-amplifier types. Typical input/output parameters for a preamplifier are noted in Fig. 7-5. Four inputs are provided, to accommodate a low-impedance microphone, a tape player, a phono player, and an FM tuner. Each input port is rated for a different signal level. Thus, the

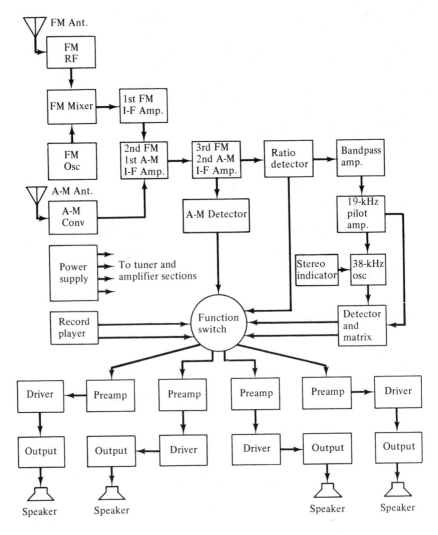

Figure 7-3 Block diagram of a hi-fi stereo-quad system.

rated input signal level for the low-impedance microphone port is 350 μV, whereas the rated level for the FM tuner port is 250 mV, or a range of 700 to 1. Observe also that *the frequency response of the amplifier is different for each input port.* The maximum available gain (MAG) in terms of voltage for the exemplified preamplifier is 69 dB, and an output of 1 V rms is normally delivered to a 10,000-ohm load,

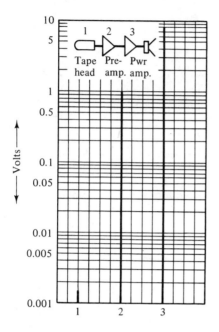

Figure 7-4   Signal-voltage levels in a typical high-fidelity channel.

regardless of the input port that is utilized. Next, consider the typical voltage-gain and power-output values for a power amplifier as noted in Fig. 7-6. A signal-voltage gain of 18 dB is provided, and an input of 1 V rms produces approximately 8 V rms across an 8- or 16-ohm load. Since the input impedance of the power amplifier is 10,000 ohms, its power gain is considerable, although its voltage gain is only moderate.

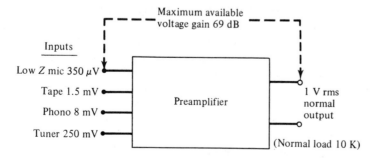

Figure 7-5   Typical voltage gain values for a hi-fi preamplifier.

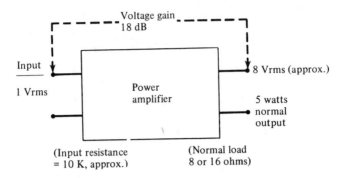

Figure 7-6 Typical voltage-gain and power-output values for a hi-fi audio power amplifier.

7-3 BASIC AMPLIFIER CHARACTERISTICS

Note the basic amplifier characteristics for small-signal bipolar transistors shown in Fig. 7-7. The three basic configurations are characterized for a load resistance of 10,000 ohms, and a generator internal (source) resistance of 1000 ohms. Observe that the common-base (CB) configuration has the highest voltage gain, the common-collector (CC) configuration has the highest current gain, the common-emitter (CE) configuration has the highest power gain, the CC configuration has the highest input resistance, and the CB configuration has the highest output resistance. Note that *these input-resistance and output-resistance values denote AC resistance—not DC resistance.* An AC resistance is also called a dynamic resistance, or an incremental resistance. Inasmuch as the CE configuration has the least difference between its input- and output-resistance values, it is the most commonly used audio-amplifier configuration.

Observe that *the power gain of a stage is equal to the product of its voltage gain and current gain.* Although the CB configuration has a very large difference between its input- and output-resistance values, this relation is occasionally useful to the designer for impedance matching in circuits or systems. Inasmuch as the CC configuration has the lowest output resistance, it is useful as an impedance transformer when the designer needs to supply substantial signal power to a comparatively low-resistance load from a high-impedance source. As seen in Fig. 7-8, the characteristics of a CE amplifier configuration vary considerably in some respects as the collector load resistance is varied in value. As seen in Fig. 7-9, the power gain in a particular configuration varies substantially with changes in load resistance from 10 ohms to 1 megohm.

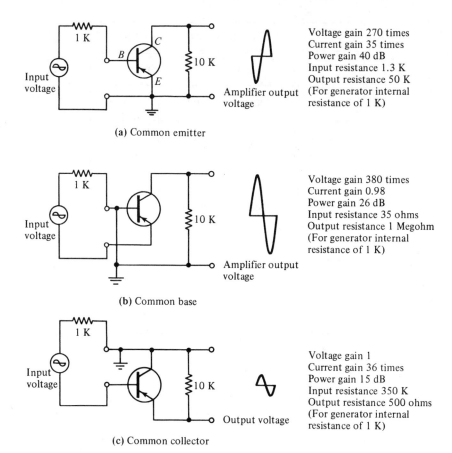

Voltage gain 270 times
Current gain 35 times
Power gain 40 dB
Input resistance 1.3 K
Output resistance 50 K
(For generator internal
resistance of 1 K)

Amplifier output
voltage

(a) Common emitter

Voltage gain 380 times
Current gain 0.98
Power gain 26 dB
Input resistance 35 ohms
Output resistance 1 Megohm
(For generator internal
resistance of 1 K)

Amplifier output
voltage

(b) Common base

Voltage gain 1
Current gain 36 times
Power gain 15 dB
Input resistance 350 K
Output resistance 500 ohms
(For generator internal
resistance of 1 K)

Output voltage

(c) Common collector

Figure 7-7 Basic amplifier characteristics for small-signal bipolar transistors.

The 10,000-ohm value on which the data in Fig. 7-7 are based is a typical bogie (design-center) value for the output resistance of a preamplifier and for the input resistance of a power amplifier. Observe that the maximum power gain (MPG) for the CE configuration occurs for a load resistance of approximately 50,000 ohms. *Small-signal operation denotes an input level in the range from 1 μV to 10 mV.*

7-4 COMPONENT AND DEVICE TOLERANCES

Components and devices have exact values and characteristics only in theory. In practice, every fabricated unit is subject to tolerances; in a

	Common-emitter configuration		
	Collector load resistance		
Parameter	1 K	10 K	100 K
Voltage gain	30	270	1,000
Current gain	50	35	20
Power gain	30 dB	40 dB	43 dB
Input resistance	1.3 K	1.3 K	1.2 K
Output resistance	60 K	50 K	40 K

(For generator internal resistance of 1 K).

Figure 7-8 Amplifier characteristics versus collector load resistance value.

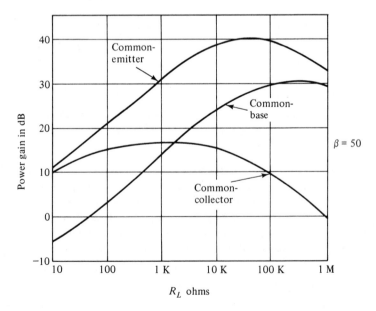

Figure 7-9 Power gain versus load resistance value for a small-signal bipolar transistor.

system, these tolerances can become of primary concern to the designer. At this point, it is helpful to consider the effect of component and device tolerances on signal power output from the CE configuration. Refer to Fig. 7-9. It is seen that a ±20 percent tolerance on a 10,000-ohm load resistor will not result in a large tolerance on signal-power output. On the other hand, the tolerance on the current-gain (beta) value of the transistor results in a large tolerance on signal-power output. As noted previously, the power gain for an amplifier stage is equal to the product of its voltage gain and current gain. For example, the power gain is proportional to the square of current, and is proportional to the square of voltage. Therefore, a doubling of the current gain, accompanied by a doubling of the voltage gain in a stage, results in a quadrupling of the power gain. Suppose that the transistor in a stage has a beta value of 52.5 ±17.5 percent. In turn, its beta value may range from 35 to 70. If it is assumed that the stage operates linearly, a doubling in device current gain is accompanied by a doubling of stage voltage gain. Consequently, the power gain of the stage is quadrupled. In other words, *a tolerance of ±17.5 percent on the transistor beta value results in a tolerance of ±60 percent on the stage power output.*

7-5   INPUT RESISTANCE VERSUS LOAD RESISTANCE

As indicated in Fig. 7-10, the input resistance of a CE stage is almost independent of the load-resistance value. On the other hand, the input resistance of a CB stage or of a CC stage varies greatly from small to large values of load resistance. Note that the horizontal axis in the diagram is basically numbered in R_L/r_c units, where R_L is the load-resistance value and r_c is the "collector resistance" of the transistor. This latter value is very high, in the order of 1 or more megohms. If a transistor has an average beta value, of 50 for example, the horizontal axis may be alternatively numbered in R_L units, as exemplified in Fig. 7-10. The input resistance of a transistor is a function of frequency. As an illustration, a typical germanium *pnp* alloy drift-field transistor that has an input resistance of 1350 ohms at low frequencies, and up to 1.5 MHz, exhibits an input resistance of only 150 ohms at 12.5 MHz.

7-6   OUTPUT RESISTANCE VERSUS GENERATOR RESISTANCE

As shown in Fig. 7-11, the output resistance of a CE stage is reasonably independent of the generator (source) resistance value. However, *the output resistance of a CB stage or of a CC stage varies greatly from*

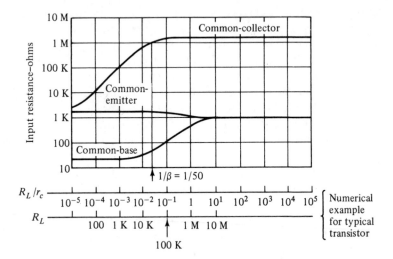

Figure 7-10   Input resistance versus load resistance for a small-signal bipolar transistor with a beta value of 50.

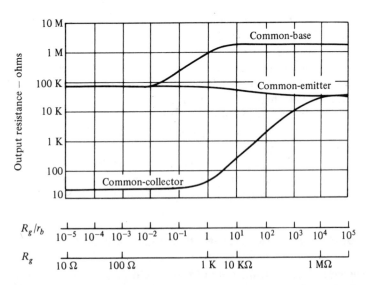

Figure 7-11   Output resistance versus generator resistance for a small-signal bipolar transistor.

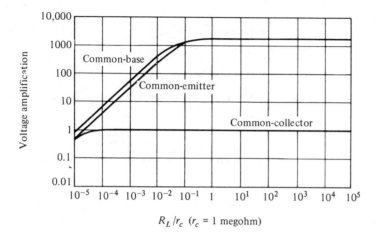

Figure 7-12   Voltage amplification versus load resistance for a small-signal transistor.

small to large values of generator resistance. Observe that the horizontal axis in the diagram is basically numbered in R_g/r_b units, where R_g is the generator resistance value, and r_b is the "base resistance" of the transistor. If it is assumed that the transistor has average characteristics, the horizontal axis may be alternatively numbered in R_g units, as exemplified in Fig. 7-11. Note that the output resistance of various transistors is also a function of frequency. For instance, a transistor that has an output resistance of 70,000 ohms at low frequencies and up to 1.5 MHz, exhibits an output resistance of only 4000 ohms at 12.5 MHz.

7-7 VOLTAGE AND CURRENT AMPLIFICATION

A stage operating in the CE, CB, or CC mode with a high value of load resistance attains a plateau in voltage amplification, as seen in Fig. 7-12. At lower values of load resistance, however, the voltage amplification decreases. Observe that the CE and CB configurations have similar trends in voltage amplification versus load resistance, although the gain of a CB stage is slightly higher than that of a CE stage at lower values of load resistance. As would be anticipated, a CC configuration has a voltage amplification of practically unity for all except very low values of load resistance. At very low values, the load approaches a short-circuit condition, and the voltage amplification trends to zero. Note in

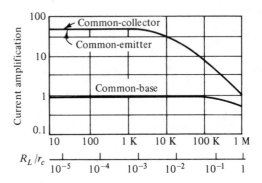

Figure 7-13   Current amplification versus load resistance for a small-signal bipolar transistor.

passing that uniformity of voltage amplification does *not* correspond to uniformity of power amplification in any of the three operating modes. This lack of correspondence results from the fact that the current amplification of a stage varies to the greatest extent over the load-resistance range where the voltage amplification varies least.

Variation of current amplification versus load resistance for the CE, CB, and CC modes is shown in Fig. 7-13. In contrast to the voltage-amplification trend, the current amplification of a stage attains a plateau at low values of load resistance. Because *the power amplification of a stage is equal to the product of its voltage amplification and its current amplification,* the power output attains a maximum value when a suitable value of load resistance is utilized, as was seen in Fig. 7-9. Different values of load resistance are required to obtain maximum power amplification in the CE, CB, and CC modes of operation. In many design projects, the chief objective is to obtain maximum power output.

Both the voltage amplification and the current amplification of a stage will vary in accordance with the *tolerance on the beta value* of the transistor. On the other hand, tolerances on the load-resistance value have much less effect on the values of voltage and current amplification. Transistors in most production lots have a comparatively wide tolerance on beta value. Accordingly, the designer ordinarily chooses somewhat elaborated amplifier circuitry that tends to minimize the effect of tolerances on the beta value. There are some design situations in which a pair of transistors in a stage must have beta values that are practically the same, if the stage is to perform at low distortion. Selected pairs of transistors are employed in this situation; they are called *matched pairs.*

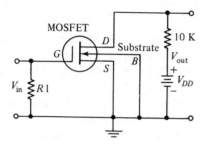

Voltage gain: 50 times
Transconductance: 5000 μmhos
Power gain: 17 dB (50 times)
Input resistance: very high
Output resistance: 20 K
(For generator internal resistance
of 500 ohms)

(a) Common source

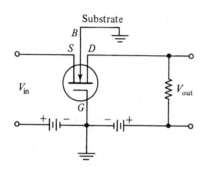

Voltage gain: 1.8
Input resistance: 240 ohms
Output resistance: High
(For generator internal resistance
of 500 ohms)

(b) Common gate

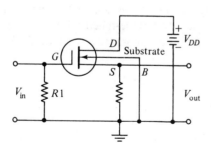

Voltage gain: 0.5
Input resistance: 2 M
Output resistance: 240 ohms
(For generator input resistance
of 500 ohms)

(c) Common drain

Figure 7-14   Basic FET amplifier configurations and characteristics.

7-8   BASIC FET AMPLIFIER CHARACTERISTICS

Field-effect transistors (FET's) are different from bipolar transistors chiefly in that an FET gate is isolated in the device structure, and does not draw current. Two basic classes of FET's are called junction field-effect transistors (JFET's) and insulated-gate field-effect transistors (IGFET's). An IGFET has a junction-type gate and will draw gate current if it is forward-biased; this device is similar to a triode electron tube. Three basic FET amplifier configurations are utilized by the designer, as shown in Fig. 7-14. These examples depict the metal-oxide-semiconductor field-effect transistor (MOSFET). The common-source mode is in widest use. Since the MOSFET gate draws negligible current, it is termed a *voltage-operated device,* whereas a bipolar transistor is called a *current-operated device.* These are only relative terms, but they indicate a basic distinction. Thus, the gain of an FET is stated in *transconductance* units, whereas the gain of a bipolar transistor is stated as a *current ratio* (beta).

The FET depicted in Fig. 7-14 has a transconductance (Gm) value of 5000 μmhos. Accordingly, if the stage utilizes a 10,000-ohm load resistor, the stage voltage gain will be 50 times. Note that the power gain is 17 dB, which is considerably less than the 40 dB provided by a corresponding CE bipolar transistor amplifier stage. Comparative design characteristics are listed in Chart 7-1. The circuit designer will often choose an FET stage instead of a bipolar transistor stage when very high input resistance is required. Note that the output resistance of an FET stage is substantially less than that of a corresponding bipolar transistor amplifier stage. In many audio design situations, this is a secondary consideration. There are two basic classes of FET's. The *depletion* type draws substantial current from V_{DD} when there is no input signal present. On the other hand, the *enhancement* type of FET draws practically no current from V_{DD} when no input signal is applied. Audio amplifiers are usually designed with depletion-type FET's. Enhancement-type FET's are ordinarily used in digital-logic circuitry.

7-9   PRINCIPLES OF NEGATIVE FEEDBACK

To minimize amplifier distortion by reduction of inherent nonlinearity in transistor characteristics, the designer utilizes negative-feedback action. The basic principle of negative feedback operation is shown in Fig. 7-15. A portion of the output signal is fed back and mixed with the input signal so that partial cancellation of the source voltage occurs.

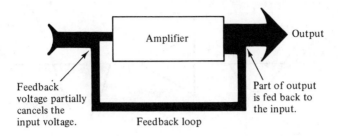

Figure 7-15 Basic principle of negative feedback.

An extreme example of nonlinear distortion is depicted in Fig. 7-16, to illustrate the process of partial cancellation at the input. Note that the mixture of the source waveform with the negative-feedback waveform serves to predistort the input signal to the transistor. This predistortion tends to compensate for the distortion introduced by the transistor, so that the amplifier output signal has reduced distortion. If a sufficiently large amount of negative feedback is utilized by the designer, an amplifier can be made as nearly distortionless as desired. Observe that negative feedback does not make a transistor operate linearly—*negative feed-*

CHART 7-1

Comparative Device Parameters

Parameter	Bipolar Transistor	JFET Transistor	MOSFET Transistor
Input impedance	Low	High	Very high
Noise	Low	Low	Unpredictable
Aging	Not noticeable	Not noticeable	Noticeable
Bias voltage temperature coefficient	Low and predictable	Low and predictable	High and unpredictable
Control electrode current	High	0.1 nA	10 pA
Overload capability	Comparatively good	Comparatively good	Poor

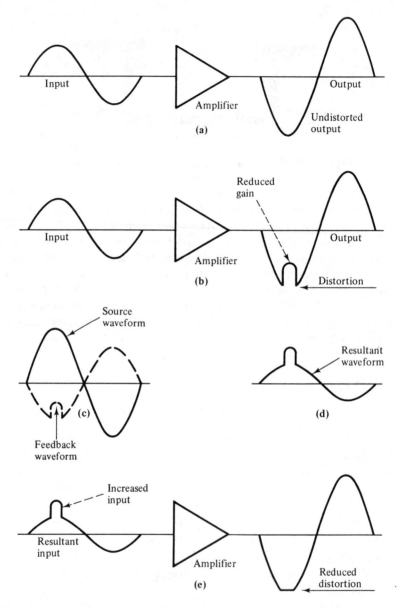

Figure 7-16   Example of amplifier operation with and without negative feedback. **(a)** Ideal amplifier operation with undistorted output; **(b)** amplifier operation with serious distortion present; **(c)** source waveform mixed with negative-feedback waveform; **(d)** resultant input waveform to amplifier; **(e)** reduced distortion in output waveform owing to negative feedback.

back merely compensates for nonlinearity introduced by the transistor, so that the overall amplifier operation is effectively linearized.

Refer to Fig. 7-17. This diagram illustrates the circuit action in the negative-feedback amplifier arrangement shown in Fig. 7-16. Since a 10-mV input to the amplifier provides a 2-V output, the gain of the amplifier by itself is 200 times. Next, observe that the output voltage drives the negative-feedback circuit. The output from the feedback network is 90 mV. This feedback voltage opposes the source voltage of 100 mV, leaving 10 mV at the input terminals of the amplifier. Accordingly, the end result is that a generator voltage of 100 mV produces 2 V output from the amplifier. That is, the gain of the stage is 20 times. Negative feedback in this example has reduced the stage gain from 200 to 20 times. Although this is a large reduction in gain, it represents an essential *tradeoff* to minimize distortion.

Another advantage of negative feedback is that tolerances on transistors have less effect on amplifier action than if negative feedback were not used. For example, suppose that the transistor in Fig. 7-17 is replaced by another transistor that provides only one-half of the gain of the original transistor. If negative feedback were not used, the output voltage would drop to one-half of its original value. On the other hand, when the foregoing negative-feedback arrangement is employed, the output voltage drops to 90 percent of its former value. In other words, this *negative-feedback action has reduced a 50 percent tolerance on the transistor beta value to an effective tolerance of 10 percent.* In many design situations, this is a desirable tradeoff. As an illustration, it helps to ensure that a production run of amplifiers will have a specified

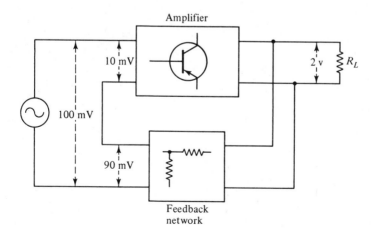

Figure 7-17 Example of negative-feedback circuit relations.

minimum gain value, regardless of wide tolerances on beta values of production-lot transistors.

Negative feedback also reduces the noise that is produced within the amplifier itself. As an illustration, suppose that the transistor depicted in Fig. 7-17 produces noise pulses as a result of marginal collector-junction leakage. A noise pulse then appears at the output of the stage; simultaneously, a cancellation pulse is fed back to the input of the transistor, thereby greatly reducing the disturbance in the output. This noise-reduction process takes place in the general manner depicted in Fig. 7-16. Observe that the negative-feedback arrangement in Fig. 7-17 is said to have *significant feedback*. This means that sufficient negative feedback is employed to reduce the gain of the amplifier to 25 percent or less of its original value. In this example, the amplifier gain is reduced to 10 percent of its original value. *A voltage gain reduction to 10 percent is also called 20 dB of negative feedback.*

Negative feedback also improves the frequency response of an amplifier. As an illustration, suppose that the transistor in Fig. 7-17 has a beta cutoff frequency of 10 kHz. In turn, if no negative feedback were utilized the frequency response of the amplifier would be as shown by the solid curve in Fig. 7-18. On the other hand, if 20 dB of negative feedback is employed, the frequency response of the amplifier is improved as shown by the dashed curve in Fig. 7-18. Thus, *a tradeoff of*

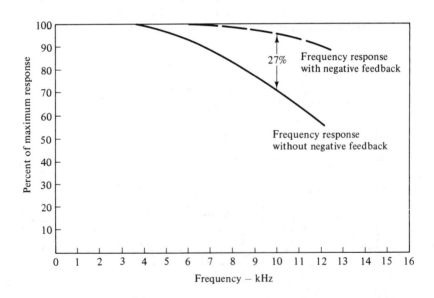

Figure 7-18 Negative feedback improves amplifier frequency response.

20 dB in gain provides a 27 percent improvement in gain of the transistor at its −3-dB cutoff point. If the designer uses less negative feedback, the improvement in frequency response will be less; if he employs more negative feedback, the improvement in amplifier response will be greater than in the example cited.

There are two basic negative-feedback arrangements, termed *voltage feedback* and *current feedback,* as shown in Fig. 7-19. Voltage feedback is provided by a resistor connected from collector to base of the transistor. Current feedback is provided by a resistor connected in series with the emitter lead. Note that voltage feedback decreases the output resistance of an amplifier. On the other hand, current feedback increases the output resistance of an amplifier. Both voltage feedback

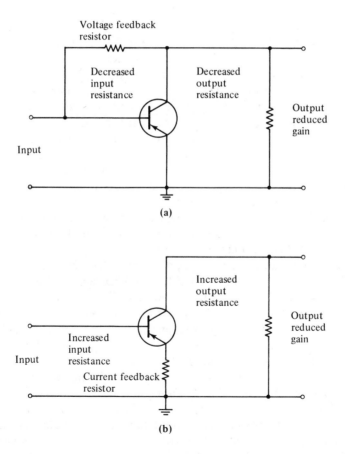

(a)

(b)

Figure 7-19 Basic negative-feedback arrangements. **(a)** Voltage feedback; **(b)** current feedback.

and current feedback operate by opposing the input voltage to the amplifier by a certain fraction of the output voltage. The fraction of the output voltage that is mixed (in phase opposition) with the input voltage is often termed B. The amplification of the stage without negative feedback is often termed A. These terms are useful to calculate the approximate reduction in harmonic distortion that results from use of negative feedback, in accordance with the equation

$$D_n = \frac{D_o}{1 + AB}$$

where D_n is the percentage distortion with negative feedback.

D_o is the percentage distortion without negative feedback.

A and B are defined as noted above.

As an illustration of harmonic distortion reduction by means of negative feedback, suppose that the amplifier depicted in Fig. 7-17 develops 5 percent harmonic distortion without negative feedback. In this example, $A = 200$, and $B = 0.45$. Accordingly, the percentage of distortion with negative feedback will be approximately 0.5 percent. In other words, *20 dB of negative feedback provides a 90 percent reduction in harmonic distortion in this situation.* Observe in Fig. 7-17 that a generator voltage of 100 mV results in an output voltage of 2V from the stage. If no negative feedback is used, a generator voltage of 10 mV will result in an output of 2 V. Thus, when negative feedback is employed, the input signal may be increased to obtain the same output level that would prevail with no negative feedback. On the other hand, if the input signal level cannot be increased, or cannot be increased sufficiently to obtain the required output level, the designer must cascade stages as may be necessary.

In addition to the advantages of negative feedback that have been noted, the tolerance on the value of load resistance is also relaxed. With reference to Fig. 7-17, suppose that no negative feedback is employed, and that the value of R_L is increased 20 percent. From a practical viewpoint, the output voltage will increase 20 percent as a result. In other words, the output level will increase from 2 V to 2.4 V. On the other hand, when 20 dB of negative feedback is utilized, and the value of R_L is increased 20 percent, the output level will increase by only 1.5 percent, to a level of 2.03 V, approximately. Still another advantage of negative feedback is the effective stabilization of beta value that it provides for a transistor under conditions of temperature variation. *The value of beta increases approximately 3 to 1 in the temperature range from $-55°$ to $+85°C$;* this is equivalent to a ±50 percent tolerance on the beta value. However, if the designer uses 20 dB of negative feedback, this variation is reduced to an effective tolerance of ±10 percent on the beta value.

In addition to the extension of high-frequency response that results from the use of conventional negative feedback, additional extension can be provided by the use of a frequency-compensated negative-feedback loop, as shown in Fig. 7-20(a). Capacitor C has less

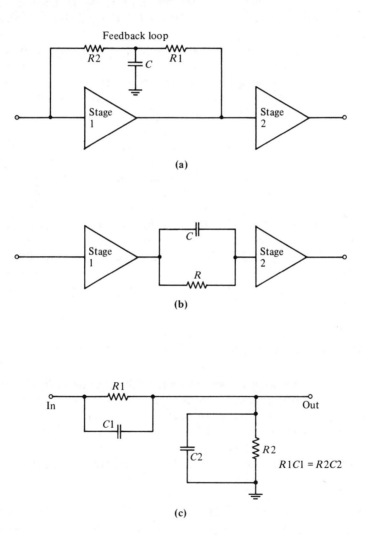

(a)

(b)

(c)

Figure 7-20 Frequency-selective and frequency-independent RC circuits. (a) Frequency-selective negative-feedback loop; (b) frequency-selective amplifier-coupling circuit; (c) frequency-independent RC voltage divider.

reactance as the frequency is increased. In turn, the amount of negative feedback decreases at higher frequencies, and the amplifier gain increases. There is a tradeoff involved in this arrangement, because the percentage of harmonic distortion will increase at higher frequencies. A limit on the amount of extension in high-frequency response is imposed by the cutoff frequency characteristic of the transistor that is utilized. That is, *the transistor beta value decreases to unity at some upper frequency limit;* above this frequency, the transistor necessarily imposes a loss instead of developing gain.

An alternative form of frequency-selective circuitry is shown in Fig. 7-20(b). Note that this is not a feedback loop; it is an amplifier coupling circuit. Its frequency characteristics depend upon the input capacitance and resistance of the second stage. When *R* and *C* have suitable values, an extension of the amplifier's high-frequency response is obtained, owing to the decreasing reactance of *C* with increasing frequency. A tradeoff is also involved in this arrangement, inasmuch as the *RC* coupling network reduces the amplifier gain at lower frequencies. Next, observe the frequency-independent *RC* voltage divider depicted in Fig. 7-20(c). This configuration is used in an amplifier network to reduce the signal level without introducing any frequency distortion. In other words, if the time constant of the first section is equal to the time constant of the second section, the output voltage will be a specified fraction of the input voltage at any frequency. That is, if a complex waveform such as a square wave is being processed by the amplifier, no waveform distortion will be imposed by the frequency-independent *RC* voltage divider.

7-10 ACTIVE TONE CONTROL

Designers often employ active tone-control arrangements, in which *RC* filter sections with adjustable resistance elements are included in the input and/or output branches of one or more amplifier stages. A two-stage arrangement with separate bass and treble tone controls is shown in Fig. 7-21, with functional diagrams for the bass-control and treble-control sections. Bass boost results from bypassing of the higher audio frequencies by $C2$, and bass cut results from attenuation of the lower audio frequencies by $C1$. Treble boost results from action of $C4$, whereas treble cut results from action of $C5$. The frequencies at which boost and cut action starts depend upon the relative capacitor values.

In most tone-control designs, capacitor values are chosen so that the treble control has little or no effect on bass response, and vice versa. Typical frequency-response curves for extreme settings of the tone con-

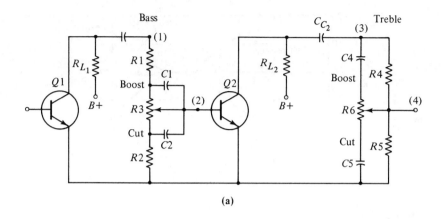

(a)

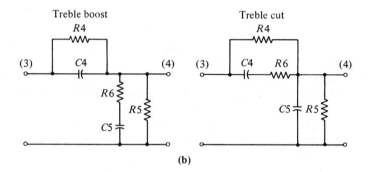

(b)

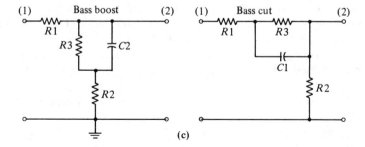

(c)

Figure 7-21 Typical configuration for an active bass-treble tone control. **(a)** Bass/treble RC tone-control sections with device isolation; **(b)** treble section equivalent circuits at extreme control settings; **(c)** bass section equivalent circuits at extreme control settings.

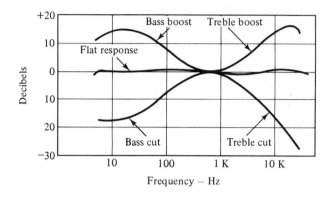

Figure 7-22 Frequency-response curves for extreme settings of a typical bass-treble tone-control arrangement.

trols are shown in Fig. 7-22. Note that although the configuration exemplified in Fig. 7-21 contains two transistors, its overall gain is about unity, owing to insertion losses of the RC filter sections. That is, *the active portion of the arrangement merely cancels out the insertion loss of the passive sections.*

An active tone-control arrangement with JFET's is depicted in Fig. 7-23. Unlike the previous configuration, this arrangement does not use device isolation between the bass and treble RC sections. Interaction between subsections is extensive. Note that $R23$ is a source-bias resistor for $Q2$. It develops a voltage drop that determines the operating point for $Q2$. Because the resistor is not bypassed, it develops negative feedback in this stage, thereby improving the fidelity (linearity) of response. Decoupling is provided by $C16/R21$. Self-bias for $Q1$ is provided by the source-bias resistors $R14$ and $R15$. Only $R15$ develops negative feedback, because $R14$ is bypassed. In turn, less gain is traded off for improved fidelity of response in the $Q1$ stage than in the $Q2$ stage.

Observe that in the absence of negative feedback, each stage in Fig. 7-23 will develop a typical gain of 40 times. However, the effect of degeneration owing to an unbypassed source resistor reduces this value substantially. Thus, if the $Q1$ stage has a gain of 40 times without degeneration, its gain will be about 8.8 times when $R15$ is unbypassed. Similarly, if the $Q2$ stage has a gain of 40 times without degeneration, its gain will be about 3.3 times when $R23$ is unbypassed. To calculate the stage gain with the source resistor bypassed, multiply the trans-conductance of the JFET by the value of the drain (load) resistor. In the case of the Q2 stage, the 47,000-ohm resistor is effectively con-

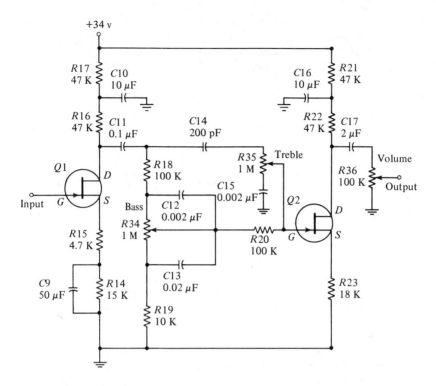

Figure 7-23 Active bass-treble tone-control arrangement with JFET's.

nected in parallel with the 100,000-ohm volume control. In turn, the designer calculates stage gain on the basis of a 32,000-ohm drain-load resistor. Design-center transconductance values for JFET's range up to 2000 μmhos, or more.

7-11 POWER AMPLIFIER PARAMETERS

There is no sharp dividing line between small-signal amplifiers and power amplifiers. Driver amplifiers occupy an intermediate position, for example, between a preamplifier and a power amplifier, or between an oscillator and an output power amplifier, as depicted in Fig. 7-24. Audio power amplifiers may operate in class A, class AB, or class B. Class AB and class B amplifiers are always operated in push-pull or equivalent configurations. *Most power amplifiers require transistor operation at power levels that are near the runaway condition.* This

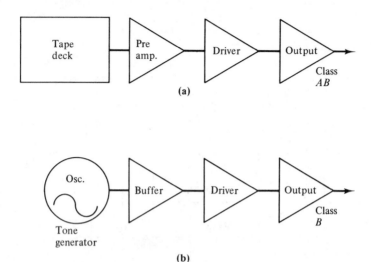

Figure 7-24 Example of power-amplifier arrangements. **(a)** Tape deck, preamp, driver, and output amplifier; **(b)** tone generator, buffer, driver, and output amplifier.

hazard is typically aggravated by the use of bias networks that have marginal stability in the interest of operating efficiency. Because thermal runaway in power-amplifier stages is likely to cause catastrophic destruction of the transistors, the designer must be guided by worst-case principles to eliminate, or at least to minimize, the possibility of thermal runaway.

Worst-case conditions involve the onset of indefinite increase in the value of beta, zero base-emitter voltage, minimal load impedance, and collector leakage current at its maximum value. In a class B power amplifier, maximum transistor power dissipation occurs at the time when the signal power output is 40 percent of its maximum value; the power dissipated by each transistor is then 20 percent of the maximum power output. On the other hand, in a class A amplifier, maximum transistor power dissipation occurs in the absence of an applied signal.

7-12 COMPLEMENTARY SYMMETRY AMPLIFIER OPERATION

There has been a marked trend to the use of *complementary symmetry design* in audio power amplifiers. Advantages of this configuration are: (1) reduced circuit complexity, (2) elimination of separate phase-

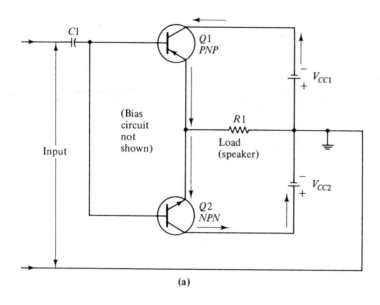

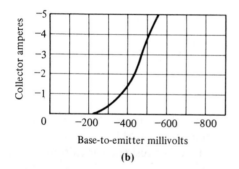

(b)

Figure 7-25 A zero-bias complementary-symmetry configuration. (a) Skeleton circuit diagram; (b) typical power-transistor transfer characteristic.

inverter stage, and (3) provision of extended frequency response, owing to reduced common-mode conduction. A basic configuration is shown in Fig. 7-25. $Q1$ and $Q2$ operate in the CC mode. Each transistor conducts over one-half of an input cycle. Resultant action in the output circuit can be understood by observing the circuit depicted in Fig. 7-26. This is a simplified version of the output arrangement. The internal

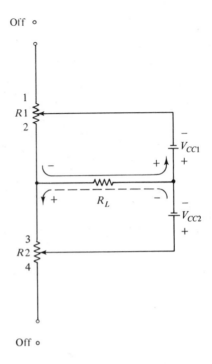

Figure 7-26   Simplified version of a complementary-symmetry output circuit.

emitter-collector circuit of $Q1$ is represented by variable resistor $R1$, and that of $Q2$ is represented by variable resistor $R2$. While one resistor is in its "off" position, the other resistor varies through its range. Then their roles are reversed. It is evident that the two transistors must have closely matched characteristics, or the output waveform will be distorted. A zero-bias class B stage tends to develop crossover distortion in consequence of nonlinear operation at low signal levels. Therefore, *the designer often includes a small forward bias on each transistor, so that they operate in class AB and cancel out crossover distortion.* A typical arrangement is depicted in Fig. 7-27.

Compound-connected transistors (Darlington pairs) are often used in audio power amplifiers. A Darlington pair provides a much higher beta value than a single transistor, and requires no associated components. Darlington pairs can be employed in single-ended amplifiers, in conventional push-pull amplifiers, or in complementary-symmetry configurations. An example of the latter arrangement is shown

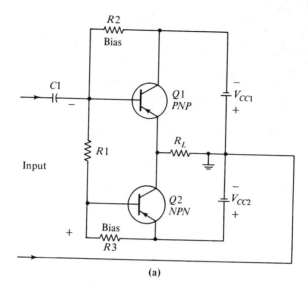

(a)

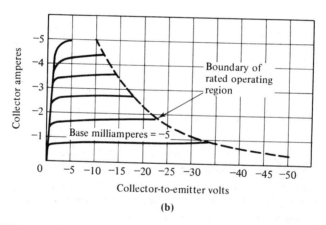

(b)

Figure 7-27 Forward-biased CE complementary-symmetry configuration. **(a)** Schematic diagram; **(b)** maximum power dissipation rating for a typical power transistor.

in Fig. 7-28. High power output is obtained, and if the Darlington pairs are well matched, distortion is low. Consequently, this configuration is in extensive use. One of its attractive features in comparison to equivalent arrangements is its simplicity and comparatively low production cost.

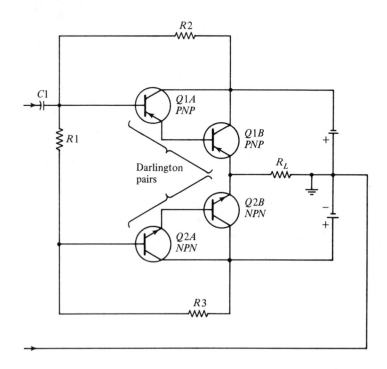

Figure 7-28 Complementary-symmetry compound-connected (Darlington) output-amplifier arrangement.

7-13 SPEAKER SYSTEM CONSIDERATIONS

It was noted previously that a speaker system includes more than one speaker. Speakers are interconnected in various ways. The simplest connection arrangement operates the speakers in series, or in parallel. A speaker has a certain value of rated input impedance, such as eight ohms. When speakers are connected in series, their total resistance increases. When speakers are operated in parallel, the system input impedance decreases. If all of the speakers in a parallel system have equal impedance ratings, the input impedance of the system is equal to the impedance of one speaker divided by the number of speakers. This type of interconnection is shown in Fig. 7-29.

When two speakers with unequal impedances are connected in parallel, the speaker with the lower impedance draws more power. For example, if a four-ohm speaker is connected in parallel with an eight-ohm speaker, *the four-ohm speaker will draw twice as much power as the eight-ohm speaker.* Accordingly, with all other things being equal,

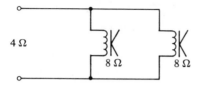

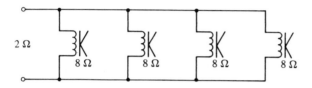

Figure 7-29 Two systems utilizing parallel-connected speakers.

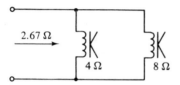

Figure 7-30 Parallel connection of speakers with different input impedances.

the four-ohm speaker will sound louder. The input impedance in this example is 2.67 ohms (Fig. 7-30). On the other hand, when speakers are connected in series, the system input impedance increases. In other words, the system input impedance will equal the sum of the impedances of the individual speakers. If the individual impedances of the series-connected speakers are not the same, the speaker that has the highest impedance will draw the most power, all other things being equal. A typical series arrangement is shown in Fig. 7-31.

Another system interconnection arrangement employs series-parallel circuitry, as exemplified in Fig. 7-32. The impedance of a series-parallel system is found by first calculating the impedance of each series "string"; then, each string is considered as an individual speaker, and the input impedance of the parallel arrangement is calculated. In practice, series-parallel systems almost always employ speakers that have the same impedance ratings. On the other hand, speakers of differ-

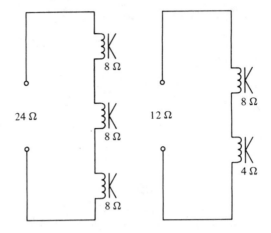

Figure 7-31   Typical series speaker arrangements.

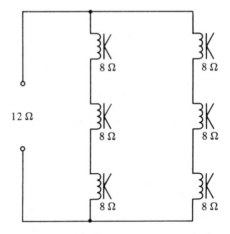

Figure 7-32   A series-parallel speaker arrangement.

ent types, such as woofers, squawkers, and tweeters, are generally operated at different power levels.

A *crossover circuit* is required when a tweeter is operated with a woofer, because a tweeter will not withstand high power at low frequencies. The simplest crossover configuration consists of a capacitor connected in series with a tweeter, as depicted in Fig. 7-33. Since capacitive reactance decreases as the frequency increases, the reactance (impedance) of the capacitor at some particular frequency will be

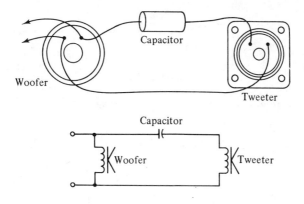

Figure 7-33 A simple crossover arrangement.

equal to the impedance of the tweeter. This is called the crossover frequency of the circuit. The value of the capacitor is chosen with respect to the input impedance of the tweeter. Consider the following example:

If a four-μF capacitor is connected in series with an eight-ohm tweeter, the crossover frequency will be approximately 5 kHz. Again, if an eight-μF capacitor is utilized, the crossover frequency will be approximately 2.5 kHz. When electrolytic capacitors are used in a crossover network, it is essential to use the *nonpolarized* type, inasmuch as the capacitor operates in an AC circuit. Note that the power delivered to the tweeter decreases to one-half, or -3 dB, at the crossover frequency in Fig. 7-33. As the operating frequency is progressively decreased, the power demand of the tweeter continues to decrease, until practically no power is drawn by the tweeter at bass frequencies. Conversely, as the operating frequency is progressively increased, the power demand of the tweeter continues to increase until the capacitor may be regarded as a short circuit at high audio frequencies.

A series capacitor prevents low frequencies from energizing the tweeter. However, this simple arrangement does not prevent high frequencies from energizing the woofer. If the woofer is capable of reproducing frequencies above the crossover value, both speakers will reproduce the midrange, and this response may tend to overemphasize the midrange frequencies. Therefore, it is general practice to connect a coil in series with the woofer, as depicted in Fig. 7-34. Note that the reactance variation of a coil versus frequency is opposite from that of a capacitor. That is, as the frequency increases, the reactance (impedance) of a coil also increases. Note that a coil with an inductance

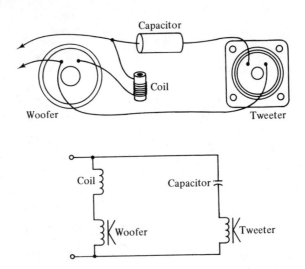

Figure 7-34   A coil and a capacitor provide a crossover network.

value of 0.25 mH has a reactance of approximately eight ohms at 5 kHz. Or, if the inductance value of the coil is increased to 0.5 mH, the crossover frequency will be approximately 2.5 kHz.

Consider the total impedance that a speaker system with a crossover network presents to the output terminals of an amplifier. For example, if eight-ohm speakers are used, as shown in Fig. 7-35, it might appear that the input impedance of the system would be four ohms. However, this is not a true analysis. Observe that the continually changing impedances of the coil and capacitor branches versus frequency will hold the system impedance at approximately eight ohms. If we assume that the speakers present a purely resistive load, *the total impedance will not vary by more than 1 percent over the audio-frequency range.* Note in the diagram that the system input impedance remains virtually eight ohms at 1/10 and at 10 times the crossover frequency.

A speaker has some inductance, as well as resistance, and both of these contribute to its input impedance at a chosen frequency. The speaker input impedance will increase with an increase in operating frequency. This makes the actual impedance values somewhat higher than the rounded-off impedance values indicated in Fig. 7-35. *If an efficient tweeter is used, it will compensate for the tendency of the woofer's increase of input impedance with an increase in operating frequency.* Observe also that when the simple crossover arrangement depicted in Fig. 7-34 is used, the rising impedance of the woofer versus

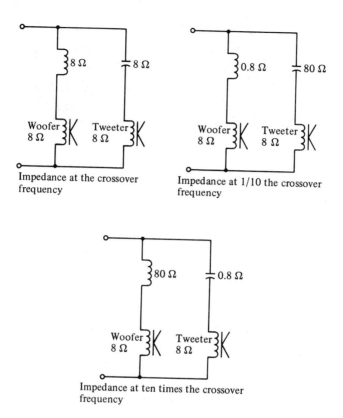

Figure 7-35 Input impedances over the audio range.

frequency assists in preventing the system input impedance from falling lower than approximately eight ohms as the operating frequency increases.

To achieve a tonal balance under various acoustic conditions, it is generally necessary to include a level control in the tweeter circuit, as shown in Fig. 7-36. Since tweeters are usually more efficient than woofers, tonal balance ordinarily requires some attenuation of the drive voltage to the tweeter. A tweeter control such as that depicted in Fig. 7-36 may be a wire-wound potentiometer with a power rating of at least two watts and a total resistance of 50 ohms. However, if a horn-type tweeter is used, the total resistance of the potentiometer should be 150 ohms. A typical speaker switching system is shown in Fig. 7-37. This is a series-parallel load arrangement, with switching facilities for operation of L and R channels individually or together. The audio driving

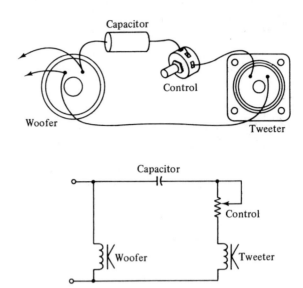

Figure 7-36   Tweeter level control.

power can also be switched to some external speaker system, if desired. This system uses two 15-inch woofers, four 3½-inch tweeters, and two horn tweeters. Nonpolarized electrolytic capacitors are utilized. Each woofer has a rated input impedance of 6.4 ohms, as does each of the horn tweeters. The cone tweeters have a rated input impedance of 4.5 ohms.

7-14   TONE-BURST TEST OF SPEAKER RESPONSE

High-fidelity speakers are usually tested by means of tone bursts. As shown in Fig. 7-38, tone-burst tests are typically made at 100 Hz, 1000 Hz, and 10,000 Hz. The sound-wave output from the speaker is picked up by a laboratory-type microphone, and fed to an oscilloscope. In turn, the design engineer evaluates the resulting distortion produced by the speaker in the tone-burst waveform. It is desirable that the leading and trailing edges of the reproduced burst waveform be nearly identical, as exemplified in the diagrams. This requires a flat frequency response and adequate damping in the speaker system.

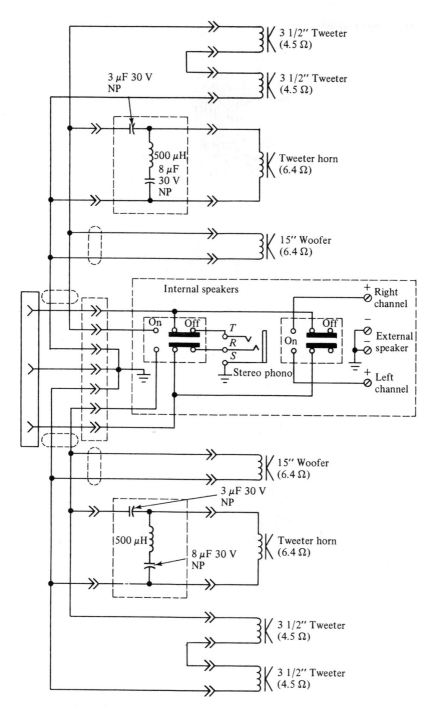

Figure 7-37   A typical L and R speaker switching system.

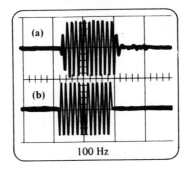

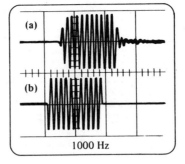

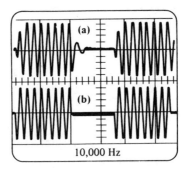

Figure 7-38 Typical speaker tone-burst tests. **(a)** Sound output waveform produced by speaker; **(b)** tone-burst waveform applied to speaker.

8

Community Antenna
Television Systems

8-1 GENERAL CONSIDERATIONS

Community antenna television (CATV) systems are used in fringe areas or strong-interference areas in order to provide high-quality reception (and often auxiliary programming) to all of the television and FM receivers that are connected into the system. A CATV system may serve an entire town or city. A skeleton block diagram for a CATV system is shown in Fig. 8-1. A head end in a CATV system includes a VHF amplifier for each antenna, and a signal combiner. In typical installations, an individual high-gain antenna is installed for each VHF and UHF channel; a separate high-gain antenna is also utilized for FM reception. A translator is a mixer that heterodynes a UHF signal frequency down to a VHF signal frequency. Translation serves to minimize signal loss on long cable runs. A trunk cable (coaxial cable) transports the signal energy from its originating site, such as on top of a mountain, to the utilization site, such as a town or city. The coaxial cable imposes a signal loss of approximately 1 dB per 100 feet. Therefore, an amplification of approximately 50 dB per mile of cable is required. A bridger, or bridging amplifier, splits the signal into several portions for supplying various feeder lines. Feeder amplifiers are also called *line extenders*. Both the main trunk amplifiers and the feeder amplifiers provide a gain up to 25 dB over a frequency range from 50 to 220 MHz. The CATV system impedance is 75 ohms throughout.

CATV amplifiers and signal splitters are usually provided with tilt controls. A tilt control is a filter of the bandpass type with an insertion

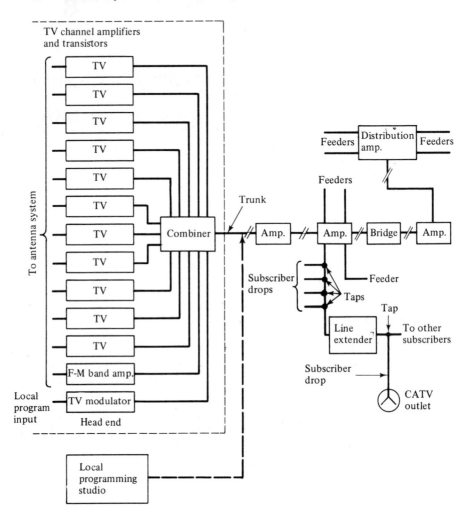

Figure 8-1   Skeleton block diagram for a CATV system.

loss that is a function of frequency. Note that the high end of the band is usually attenuated to some extent, because coaxial cable has higher losses at higher frequencies. Technically, distribution cables are tapped off from feeder cables, but both types are often termed distribution cables. A tap point along the cable is called a *subscriber tap*. Observe that a trunk line continues through a bridging amplifier. However, the last amplifier on a trunk cable is termed a *distribution amplifier*. A feeder line can be run up to 1000 feet from a bridger, before another amplifier (line extender) is required. Note in Fig. 8-1 that a subscriber drop is a cable that connects to a tap point along a feeder line. A small capacitor may be used by the designer for isolation, because it introduces some tilt. When tilt compensation is not required, an isolating resistor may be used. Distribution amplifiers must operate over a wide frequency range; they may be designed as separate low-band and high-band amplifiers, with a low-band tilt control, and supply their outputs to a single line, as exemplified in Fig. 8-2. Class A amplification is necessary, inasmuch as any amplitude nonlinearity will produce cross-modulation and mutual interference among the various signals. The signal level of each channel on the cable is on the order of 1500 μV.

8-2   FUNDAMENTALS OF RF CIRCUIT DESIGN

Most radio-frequency circuits are designed to operate in a series-resonant or parallel-resonant mode. Resonance provides controlled selectivity, and greatly increases the operating efficiency of an RF amplifier. Operating efficiency is enhanced because any active device has junction capacitances, and all circuit arrangements have stray capacitance. These capacitances become a dominant consideration at radio frequencies, because capacitive reactance is inversely proportional to the operating frequency. Efficiency is further enhanced in various design situations by employing tapped circuits to match the output impedance of one device to the input impedance of a following device. Maximum power is transferred from a source to a sink when their impedances are matched; note, however, that system efficiency is only 50 percent at maximum power transfer, as shown in Fig. 8-3. Because signal power is of far greater concern than supply power, this tradeoff is always accepted by the CATV designer.

Resonant circuits avoid the bypassing action of stray capacitance and junction capacitance by exploiting these capacitances in the tuning of inductors to resonance at the operating frequency. Here, it is instructive to consider the effect of tolerances on source and load impedances with respect to maximum power transfer. An inspection of the power variation in Fig. 8-3(a) shows that mismatching is not a highly objectionable design consideration. That is, a mismatch of ± 40 percent reduces the signal power transfer less than 10 percent. On the other hand, if the load value is low and departs 80 percent from optimum value, then 40 percent of the available signal power will be lost. However, if the load value is high, and departs 80 percent from its optimum value, the resulting loss is much less, and over 90 percent of the available signal power is still delivered to the load.

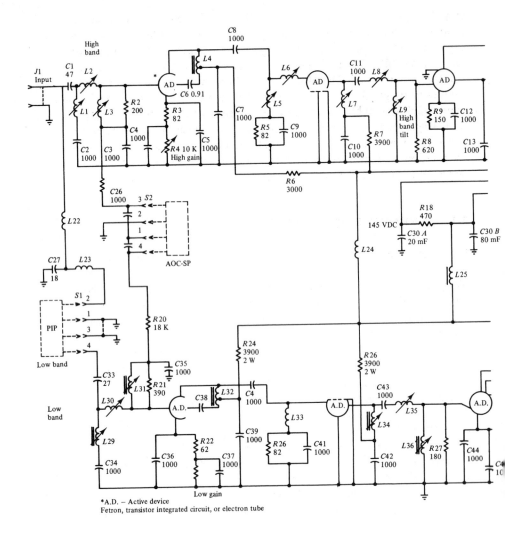

Figure 8-2   Configuration of a CATV distribution amplifier.

8-3   PRINCIPLES OF RESONANT CIRCUIT DESIGN

With reference to the four forms of parallel-resonant circuits shown in Fig. 8-4, the basic resonant-frequency equation is

$$f_r = \frac{1}{2\pi\sqrt{LC}}$$

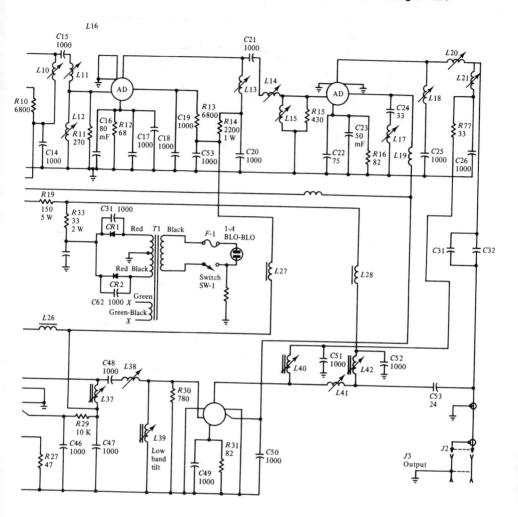

Figure 8-2 Continued

This resonant-frequency equation is exact for a resistanceless LC parallel configuration, and it is a good approximation for an LCR parallel-resonant configuration. Next, consider the effect of component tolerances on the resonant-frequency value. If the tolerance of L and C values is ± 20 percent, the worst-case condition occurs when both components have their low-limit values. That is, the resonant frequency is shifted 25 percent above its bogie value, as seen in Fig. 8-5. The quality factor, or Q, of a parallel-resonant circuit is equal to the ratio

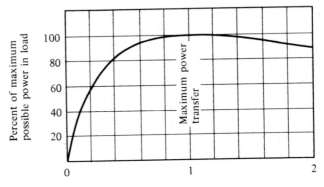

Ratio of load resistance to source resistance

(a)

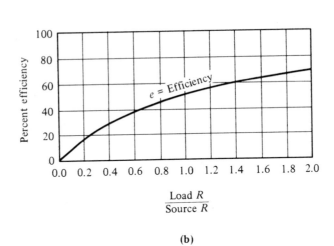

Load R / Source R

(b)

Figure 8-3 Power transfer and efficiency relations for source and load. **(a)** Power transfer versus impedance ratio; **(b)** system efficiency versus impedance ratio.

of tank current I_L or I_C to the line current, as depicted in Fig. 8-4. Note that this relationship is not true for off-resonance operation.

The circuit designer should note that if the Q value of a parallel-resonant circuit is high, that the line current I will be small, whereas the capacitor current I_C and the inductor current I_L will be large. These large currents are called the circulating currents; they merely surge back and forth between the inductor and the capacitor. Proper operation

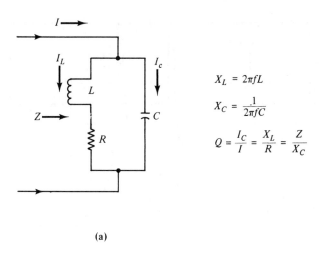

$$X_L = 2\pi f L$$

$$X_C = \frac{1}{2\pi f C}$$

$$Q = \frac{I_C}{I} = \frac{X_L}{R} = \frac{Z}{X_C}$$

(a)

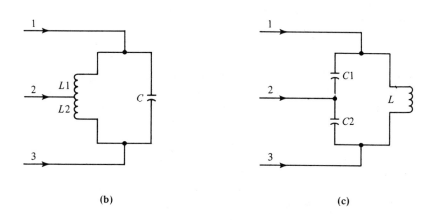

(b)

(c)

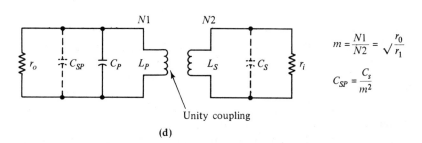

$$m = \frac{N1}{N2} = \sqrt{\frac{r_0}{r_1}}$$

$$C_{SP} = \frac{C_s}{m^2}$$

Unity coupling

(d)

Figure 8-4 Fundamental relations in basic parallel-resonant circuits.

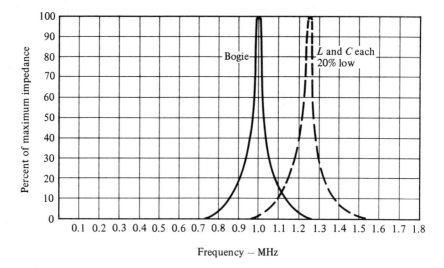

Figure 8-5 Error in resonant frequency, owing to 20 percent tolerances on *L* and *C*.

can be obtained only if the designer provides an inductor that can
handle the large circulating RF current without undue loss. Similarly,
the inductor must be connected to the capacitor with leads that have low
RF resistance. The capacitor generally imposes less of a design problem; most capacitors have a low dissipation factor and can handle comparatively large RF currents satisfactorily. Note that the *Q* value of a
component is equal to its reactance (at a specified frequency), divided
by its RF resistance. Conversely, the dissipation factor (*D*) of a component is equal to the reciprocal of its *Q* value.

To measure the RF resistance of an inductor at a specified operating frequency, the designer uses a bridge to determine its *Q* value and
its inductance value. In turn, the RF resistance of the inductor is equal
to its reactance at the specified frequency divided by its Q value. This
RF resistance value will always be greater than the DC resistance value
of the inductor, and may be very much higher at high operating frequencies. Skin effect, proximity effect, possible eddy currents, dielectric
losses, and radiation all contribute to the total value of RF resistance.
Note that the *Q* value of a parallel-resonant circuit is a measure of its
selectivity. That is, a parallel-resonant circuit has a bandpass value
equal to the number of hertz between the half-power points on its
resonance curve. This bandwidth is related to its *Q* value and resonant
frequency by the approximate equation:

$$BW \approx \frac{f_r}{Q}$$

where BW is the bandwidth in hertz of the resonant circuit.
f_r is the resonant frequency.
Q is the quality factor of the inductor.

Observe that the Q factor of a parallel-resonant circuit would be reduced from the foregoing value in the event that RF losses in the capacitor were significant. As noted previously, however, the designer is ordinarily justified in neglecting capacitor losses. Errors in design bandwidth result from tolerances on the RF resistance of an inductor. Note that the worst-case condition occurs when the RF resistance is at its lower tolerance limit. As an illustration, if the RF resistance value is 20 percent low, the Q value will be 25 percent above bogie. Accordingly, the bandwidth will be 20 percent less than bogie.

Designers often utilize tapped inductors to control RF impedance values. Refer to Fig. 8-4(b). Observe that the resonant frequency and the unloaded Q value of the parallel-resonant circuit remain the same, whether the signal is injected across points 1 and 3, or across points 2 and 3, or across points 1 and 2. It will be apparent that the impedance between points 2 and 3 is much smaller than the impedance between points 1 and 3. This ratio depends on the square of the turns ratio between terminals 1 and 3, and terminals 2 and 3. Thus, if point 2 represents a center tap, the input impedance between points 1 and 3 is four times the impedance between points 2 and 3, or between points 1 and 2. Again, if point 2 were chosen ⅓ up from point 3, the input impedance between points 1 and 3 will be nine times the input impedance between points 2 and 3. To continue the foregoing example, the input impedance between points 1 and 3 will be 2¼ times the input impedance between points 2 and 3. Observe that an error of 20 percent in location of a center tap on an inductor results in a 56 percent worst-case impedance error.

Consider next the tapped capacitance arrangement depicted in Fig. 8-4(c). As before, the resonant frequency and the unloaded Q value of the parallel-resonant circuit remain the same, whether the signal is injected across points 1 and 3, or across points 2 and 3, or across points 1 and 2. However, the input impedance between points 2 and 3 is smaller than the input impedance between points 1 and 3. As an illustration, if $C1$ and $C2$ have equal values, the input impedance between points 1 and 3 is four times the input impedance between points 2 and 3, or between points 1 and 2. This fact is obvious, because if point 2 were connected to a center tap on L, the circuit action would

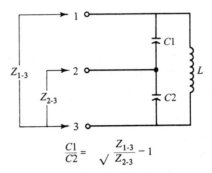

$$\frac{C1}{C2} = \sqrt{\frac{Z_{1\text{-}3}}{Z_{2\text{-}3}} - 1}$$

Figure 8-6   Relation of impedance ratios to capacitance ratios.

remain unchanged, inasmuch as a balanced bridge configuration results. The general relation versus capacitance values and input-impedance values is noted in Fig. 8-6.

In many design situations, a tuned resonant circuit (capacitor C_p and inductor L_p) in the primary of a tuned transformer is coupled to the nonresonant secondary of the transformer, as depicted in Fig. 8-4(d). In this case, if $N1$ represents the number of turns in the primary winding, and $N2$ represents the number of turns in the secondary winding, then the turns ratio m of primary to secondary under matched impedance conditions is given by the equation:

$$m = \frac{N1}{N2} = \sqrt{\frac{r_o}{r_i}}$$

where r_o and r_i are the impedance values to be matched.

If capacitance is present in the secondary circuit (C_s) of the transformer, this capacitance is reflected (or referred) to the primary circuit by transformer action as a capacitance (C_{sP}) in the primary circuit, according to the equation

$$C_{sP} = \frac{C_s}{m^2}$$

In turn, it is evident that the worst-case condition for the value of m occurs when $N1$ has its high-tolerance limit, and $N2$ has its low-tolerance limit. As an illustration, a 10 percent winding tolerance under these conditions corresponds to 22 percent error in the m value.

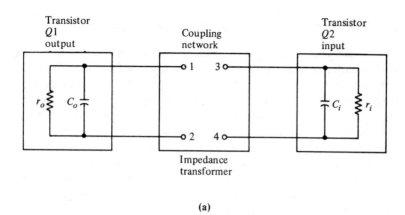

(a)

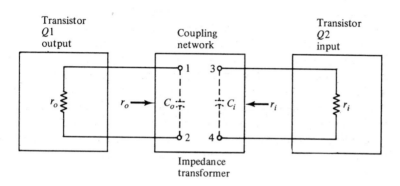

(b)

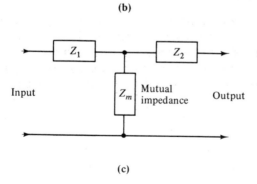

(c)

Figure 8-7 Equivalent output and input circuits for transistors with a coupling network. **(a)** Basic arrangement; **(b)** output and input capacitances; **(c)** generalized coupling network.

8-4 TRANSISTOR AND COUPLING NETWORK IMPEDANCES

The output impedance of a transistor can be regarded as a resistance r_o in parallel with a capacitance C_o, as depicted in Fig. 8-7(a). Similarly, the input impedance of a transistor can be regarded as a resistance r_i in parallel with a capacitance C_i, as indicated in the diagram. In most situations, the designer accounts for the output capacitance C_o and the input capacitance C_i by including them as part of the coupling network, as depicted in Fig. 8-7(b). Recall that there is an appreciable tolerance on the junction capacitance of a transistor, which can seldom be ignored in RF design procedures. Assume that the required capacitance between terminals 1 and 2 of the coupling network is calculated to be 500 pF. Assume also that capacitance C_o is 10 pF. A capacitor with a value of 490 pF would then be employed between terminals 1 and 2, so that the total capacitance would be 500 pF. This same method is used to compensate for capacitance C_i between terminals 3 and 4.

To cope with a production tolerance of ± 20 percent on the value of C_o, the designer could use a capacitor with a value of 488 pF, paralleled by a trimmer capacitor with a maximum value of 4 pF. Then the trimmer capacitor would be adjusted in a production test procedure to resonate the coupling network correctly. To obtain maximum power transfer from $Q1$ to $Q2$, the input impedance (terminals 1 and 2) to the coupling network must equal resistance r_o; the output impedance of the coupling network (*looking into* terminals 3 and 4) must equal r_i.

8-5 TRANSFORMER COUPLING WITH TUNED PRIMARY

A generalized representation of an inductive coupling network that consists of a single tuned circuit is depicted in Fig. 8-8(a). Capacitance C_T represents the output capacitance of transistor $Q1$, and the input capacitance of $Q2$ is referred to the primary of coupling transformer $T1$. A suitable turns ratio is chosen by the designer for $T1$ to provide a match of the $Q1$ output impedance to the $Q2$ input impedance. If the inductive reactance between terminals 1 and 2 of $T1$ is represented by L_P, then the resonant frequency f_r for this coupling transformer is given by the equation

$$f_r = \frac{1}{2\pi\sqrt{L_P C_T}}$$

Suppose that the designer utilizes the configuration in Fig. 8-8(a) for a high-frequency amplifier with two transistors operated in the *CE*

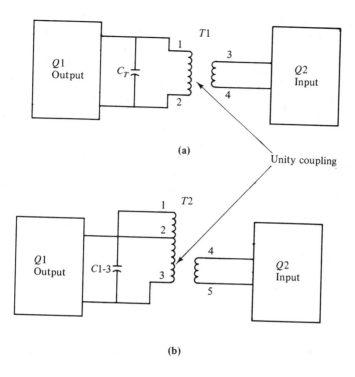

(a)

Unity coupling

(b)

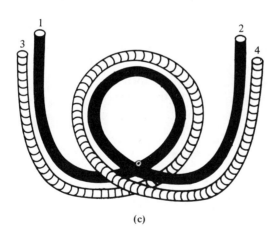

(c)

Figure 8-8   Interstage network utilizing transformer coupling with tuned primary winding. **(a)** Untapped primary; **(b)** tapped primary; **(c)** bifilar winding provides unity coupling.

mode. In turn, the basic design parameters are: resonant frequency f_r, frequency bandwidth Δf, output resistance r_o, output capacitance C_o, input resistance r_i, input capacitance C_i, coupling network insertion loss, and the power response of frequencies at the extremes of bandwidth ($f_r \pm \Delta f/2$). To obtain a sufficiently high Q value in many situations, the designer must employ a tapped primary winding, as depicted in Fig. 8-8(b). The inductance of the primary may be many times the original calculated values. As an illustration, the inductance L_{1-3} between terminals 1 and 3 in Fig. 8-8(b) could possibly be 100 times the previously calculated value for L_P between terminals 1 and 2 in Fig. 8-8(a).

To maintain the original resonant frequency, the capacitance connected across terminals 1 and 3 (C_{1-3}) must be reduced to 0.01 of the previously calculated value for C_T. To maintain a matched impedance relation for maximum power transfer, the inductance between terminals 2 and 3 of transformer $T2$ must be equal to the previously calculated value of L_P between terminals 1 and 2 of transformer $T1$. Because of production tolerances on components and devices, the designer may provide a ferromagnetic tuning slug in $L2$.

Autotransformer Coupling, Tuned Primary

Inductive coupling from the output of one transistor to the input of a second transistor can be accomplished with a tuned autotransformer, as depicted in Fig. 8-9(a). The circuit action is essentially the same as in the configuration of Fig. 8-8, except that DC isolation is not provided between transistors $Q1$ and $Q2$. Capacitance C_T in Fig. 8-9(a) includes the output capacitance of $Q1$ and the input capacitance of $Q2$, referred to the *primary section*. If L_{1-3} denotes the inductance between terminals 1 and 3 of the autotransformer, then the resonant frequency of the coupling circuit is expressed by the equation

$$f_r = \frac{1}{2\pi\sqrt{L_{1-3}C_T}}$$

The tap at terminal T on the autotransformer is specified to provide an impedance match between $Q1$ and $Q2$, as noted previously. In some applications, selectivity may be inadequate. In such a case, the Q value between terminals 1 and 3 is too low, and must be increased. This is accomplished by increasing the primary inductance; the configuration in Fig. 8-9(b) is employed for this purpose. To maintain the original resonant frequency, C_{1-4} must be reduced by the same factor that the inductance is increased, so that the product of L_{1-3} and C_T in Fig. 8-9(a) is equal to the product of L_{1-4} and C_{1-4} in Fig. 8-9(b).

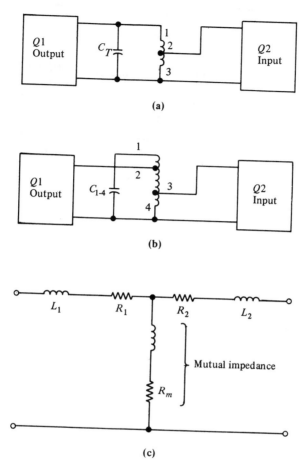

Figure 8-9 Interstage network utilizing autotransformer coupling with a tuned primary winding. **(a)** Tapped output circuit branch; **(b)** tapped output and input circuit branches; **(c)** generalized autotransformer equivalent circuit.

Impedance matching is maintained, provided that inductances $L_{2\text{-}4}$ and $L_{3\text{-}4}$ of transformer $T2$ equal the inductances $L_{1\text{-}3}$ and $L_{2\text{-}3}$, respectively, of transformer $T1$.

Capacitance Coupling

When the designer must cope with a transformer coupling problem that limits the secondary to a small number of turns, it may be impractical to employ unity coupling, or to obtain tight coupling, between primary and

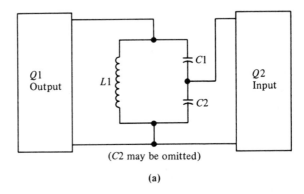

(C2 may be omitted)

(a)

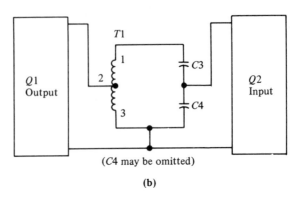

(C4 may be omitted)

(b)

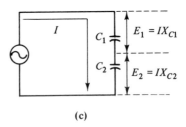

(c)

Figure 8-10 Interstage capacitance coupling with a split capacitor. (a) Untapped inductor; (b) tapped inductor; (c) capacitance voltage divider relations.

secondary. This problem is particularly troublesome in common-base configurations that operate at very high frequencies. That is, the input impedance of the transistor is less than 75 ohms in this situation. To circumvent this difficulty, the designer may employ capacitive coupling, as exemplified in Fig. 8-10. Impedance matching of the Q1 output impedance to the Q2 input impedance is obtained by selecting the proper ratio of C1 to C2. Capacitance C2 is ordinarily much larger than C1. Their reactance values bear opposite relationship to their capacitance values. In the common-base configuration, particularly, C2 may be omitted, because its reactance will be masked by the low value of input resistance.

If the designer finds that inductance L1 is too small to provide the required selectivity, he may employ the configuration shown in Fig. 8-10(b). With a tapped coil, the inductance may be increased by a chosen factor, and the total capacitance C_T reduced by the same factor in order to maintain the same resonant frequency. Impedance matching is obtained by making the inductance between terminals 2 and 3 equal to L1; the ratio of C3 to C4 must be equal to the ratio of C1 to C2. To contend with component and device tolerances in production, the designer may provide a tuning slug in the coil.

8-6 INTERSTAGE COUPLING WITH DOUBLE-TUNED NETWORKS

Advantages of double-tuned interstage coupling networks, as exemplified in Fig. 8-11, include a flatter frequency response within the pass band, a sharper drop in response immediately adjacent to the ends of the pass band, and comparatively high attenuation of frequencies not in the pass band. Observe the frequency response curves depicted in Fig. 8-12 for a double-tuned transformer with three coefficients of coupling. This example is for a resonant frequency of 1 MHz; however, the general relations apply to any operating frequency. Inductively coupled circuits with a coefficient of coupling on the order of 1 percent are often used. When two tuned coils are inductively coupled to provide bandpass filter action as in Fig. 8-12, the coupling coefficient k is given by the equation

$$k = \frac{L_m}{\sqrt{L_1 L_2}}$$

where k = coupling coefficient.
L_m = mutual inductance.
L_1 and L_2 = coil inductances.

In the present example, the double-tuned transformer employs primary and secondary coils with a Q value of approximately 100. A k

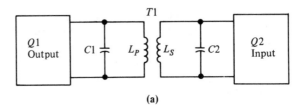

(a)

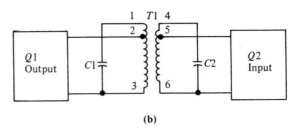

(b)

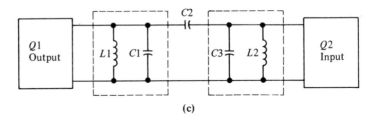

(c)

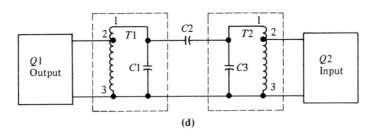

(d)

Figure 8-11   Double-tuned interstage coupling networks with inductive or capacitive coupling. **(a)** Untapped inductors; **(b)** tapped inductors; **(c)** capacitive coupling, untapped inductors; **(d)** capacitive coupling, tapped inductors.

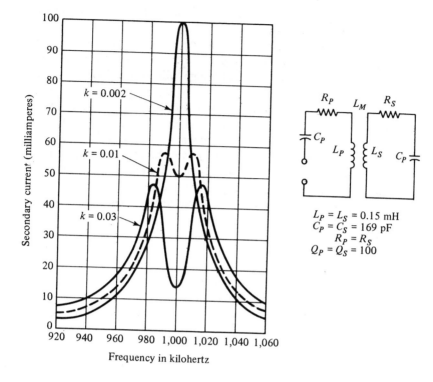

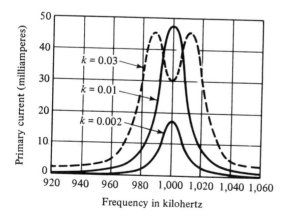

Figure 8-12 Primary and secondary characteristics for a typical double-tuned transformer, with three coefficients of coupling.

value of 1 percent provides a secondary bandwidth of 25 kHz. Greater bandwidths are provided by lower Q values. For example, the designer may shunt a resistor of suitable value across the primary and/or secondary to increase the bandwidth. Greater bandwidth can also be obtained with tighter coupling. Note that although this design increases bandwidth, it also introduces sag in the top of the response curve. Resistance loading, on the other hand, avoids excessive sag in the top of the response curve, but reduces the output signal amplitude. Note in Fig. 8-12 that the maximum secondary current occurs at critical coupling ($k = 0.002$). As shown in Fig. 8-13, greater bandwidth at higher current output could be obtained by using coils with higher Q values. However, the sag in the top of the response curve then becomes excessive. An optimum Q value results in comparatively high output and a reasonably flat-topped response curve.

Observe next the response-curve roll-off characteristics exemplified in Fig. 8-14. A single tuned coupling circuit provides comparatively poor selectivity and poor approximation to a flat-topped frequency response. However, a double-tuned coupling transformer with a kQ product of 2 provides a reasonable approximation to flat-topped peak response, and a relatively rapid drop-off in response to frequencies that

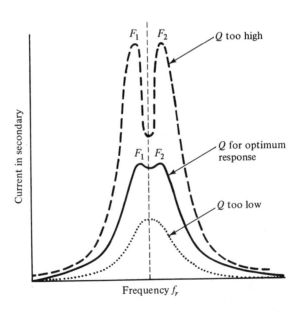

Figure 8-13 Response curves for a double-tuned transformer with three different Q values.

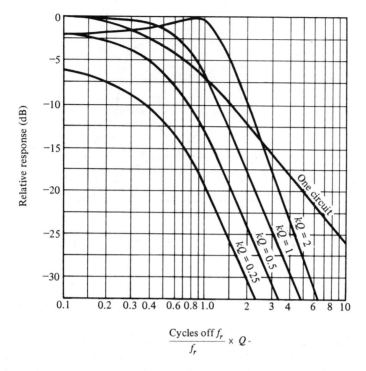

Figure 8-14 Response-curve skirt contours for single-tuned, and for three different double-tuned coupling arrangements.

are past the maximum output point. Note that kQ products less than 2 result in progressively poorer selectivity and output-level characteristics. It is generally necessary for the designer to provide tuning slugs or trimmer capacitors for the coils to contend with component and device tolerances.

Consider the inductive coupling arrangement depicted in Fig. 8-11(a), and its alternative form shown in Fig. 8-11(b). Capacitor C_1 and the primary winding L_p form a tuned circuit. Similarly, capacitor C_2 and the secondary winding L_s also form a tuned circuit. Each circuit is ordinarily tuned to the center of the pass band; the double-humped responses shown in Fig. 8-12 result from overcoupling—not from stagger tuning. However, design engineers sometimes choose a small amount of stagger tuning to supplement a moderate amount of over-coupling. Impedance matching in Fig. 8-11(a) is obtained by proper selection of the primary and secondary turns ratio. When this configuration does not provide the required selectivity, the circuit designer may

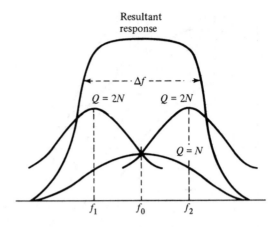

Figure 8-15 Wide-band response obtained by stagger tuning of successive stages.

use tapped primary and secondary windings, as depicted in Fig. 8-11(b).

Next, consider the requirement of wide-band amplifier response. In this situation, the designer usually employs single-tuned coupling circuits with stagger tuning, as exemplified in Fig. 8-15. Any number of stagger-tuned circuits (stages) may be utilized. If three tuned circuits are used, optimum bandwidth and response-curve characteristics are obtained when the low-frequency and the high-frequency coil have twice the Q value of the center-frequency coil. Additional bandwidth can be obtained by shunting the tuned circuits with resistance. However, the stages then develop lower gain, and additional stagger-tuned stages must then be utilized to obtain the original gain figure.

Capacitive Coupling

Note the interstage coupling networks shown in Fig. 8-11(c) and (d). Two capacitance-coupled tuned circuits are employed. With reference to Fig. 8-11(c), capacitor C_1 and inductor L_1 form a resonant circuit. Similarly, capacitor C_3 and inductor L_2 form a resonant circuit. Each coil is tuned to the same frequency. Impedance matching is obtained by properly selecting the ratio of the reactance of capacitor C_2 to the impedance of the parallel input circuit (capacitor C_3 and inductor L_2). The circuit depicted in Fig. 8-11(d) functions in the same manner as noted previously; tapped coils (autotransformers) are used to facilitate the attainment of high selectivity. To compensate with production tolerances on components and devices, the designer generally provides a tuning slug or trimmer capacitor for each inductor.

Gain Equalization

Some applications of tuned amplifiers require that the center frequency of the amplifier be variable over a wide frequency range. In this situation, either capacitance variation or inductance variation may be utilized. Ideally, a combination of both can be employed to maintain constant bandwidth and optimum frequency response over the complete tuning range. In any event, the designer is likely to encounter difficulty in maintaining constant gain over a wide frequency range, inasmuch as the transistor beta value tends to drop off more or less at the higher operating frequencies. To contend with this decrease in gain at the higher frequencies, an equalizer circuit such as that shown in Fig. 8-16 may be used. This parallel combination of R and C is designed to attenuate the low frequencies more than the high frequencies. That

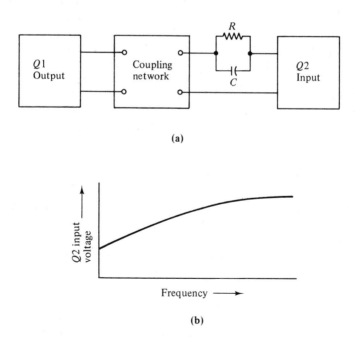

Figure 8-16 Coupled transistor amplifier with an RC equalizing circuit. **(a)** Equalizer arrangement; **(b)** typical frequency characteristic.

is, at higher operating frequencies, capacitor C bypasses a greater percentage of the signal over resistor R than it does at lower operating frequencies.

8-7   NEUTRALIZATION AND UNILATERALIZATION

A unilateral electrical device can transmit energy in one direction only. A transistor is not inherently a completely unilateral device, because voltage variations in its output circuit cause voltage variations in its input circuit to some extent. In each transistor configuration, the feedback voltage from the output aids the input voltage. Accordingly, positive feedback is present in the simpler amplifier configurations. If the positive feedback voltage is sufficiently large, the amplifier will oscillate. At radio frequencies, with tuned coupling circuits, precautions must often be taken by the amplifier designer to avoid development of regeneration or uncontrolled oscillation.

The effect of positive or negative feedback voltage on the input circuit of a device is to alter its input impedance. Usually, both the resistive and the reactive components of the input impedance are affected. This change in input impedance of a transistor owing to internal feedback can be eliminated by utilization of an external and opposing feedback circuit. If this external feedback cancels both the resistive and the reactive changes that the internal feedback produces in the input circuit, the transistor amplifier is said to be *unilateralized*. If the external feedback circuit cancels only the reactive changes produced by internal feedback, the transistor amplifier is said to be *neutralized*. That is, neutralization is a special case of unilateralization; either method will serve to prevent oscillation in an amplifier system.

Unilateralized Common-base Amplifier

A tuned common-base amplifier configuration is exemplified in Fig. 8-17. DC biasing circuits have been omitted in the interest of simplicity. Transformer $T1$ couples the input signal to the amplifier. Capacitor $C1$, with the $T1$ secondary winding, forms a parallel-resonant circuit. Transformer $T2$ couples the output of the amplifier to the following stage. Capacitor $C2$, with the $T2$ primary winding, forms a parallel-resonant circuit. The internal elements of the transistor that may cause sufficient positive feedback to cause self-oscillation are shown by dashed lines in Fig. 8–17(b). Resistor r_b' represents the resistance of the bulk material in the base region, and is called the base-spreading resistance. Capacitor C_{CB} represents the capacitance of the base-collector junction. Resistor

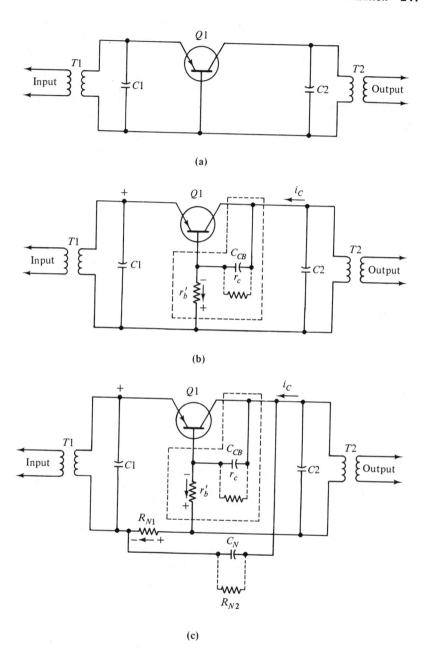

Figure 8-17   Common-base amplifier arrangements, showing internal feedback elements and external unilateralizing circuit.

r_c represents the resistance of the base-collector junction. This is a very high-valued resistance, inasmuch as the collector junction is reverse-biased. At very high operating frequencies, capacitor C_{CB} effectively shunts resistor r_c.

Assume that the incoming signal aids the forward bias voltage (causes the emitter to go more positive with respect to the base). Then collector current i_c increases in the direction shown. A portion of the collector current passes through C_{CB} and through resistor r'_b in the direction indicated, and produces a voltage drop of the indicated polarity. This voltage drop across r'_b aids the incoming signal, and accordingly develops positive feedback that may cause self-oscillation. Note in passing that, if the designer deliberately mismatches the tuned circuits to the transistor, oscillation can be prevented thereby. On the other hand, this expedient results in a loss of signal power that may not be tolerable.

When the incoming signal aids the forward bias voltage, as noted above, the collector current increases. A portion of the collector current passes through C_{CB} in the direction shown, and produces a voltage drop of indicated polarity. A portion of the collector current also passes through C_N and through R_{N1}, thereby developing a voltage drop with indicated polarity. Note that the voltages across resistors r'_b and R_{N1} are opposing voltages. If these voltages are equal, no positive or negative feedback can occur from the output circuit to the input circuit. In turn, the amplifier is said to be unilateralized.

Common-emitter Amplifier; Partial Emitter Degeneration

A common-emitter amplifier configuration that employs partial emitter degeneration is depicted in Fig. 8–18. Capacitor C_N and resistors R_{N1} and R_{N2} form a unilateralizing network. Transformer $T1$ couples the input signal to the base-emitter circuit. Resistor $R1$ forward-biases the base-emitter circuit. Capacitor $C1$ prevents short-circuiting of the base-bias voltage by the secondary winding of $T1$. Transformer $T2$ couples the output signal to the following stage. Capacitor $C2$ and the secondary winding of $T2$ form a parallel-resonant circuit. Capacitor $C3$ blocks DC battery voltage from R_{N1} and couples a portion of the collector current i_{c2} to the emitter. Inductor $L1$ is an RF choke that prevents bypassing of the AC signal to ground via $C3$.

When the input signal aids the forward bias, collector current i_c increases in the direction indicated. Internally, a portion of the collector current is coupled to the base-spreading resistance through the collector-base junction capacitance. The voltage that develops across this base-spreading resistance aids the incoming signal and develops positive

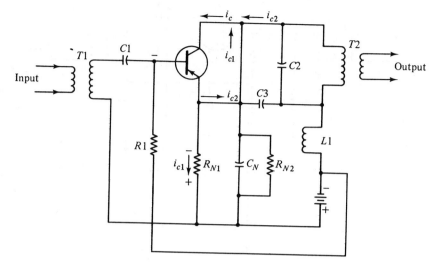

Figure 8-18 Common-emitter amplifier with partial emitter degeneration.

feedback. To offset this positive feedback, a portion of the collector current i_{c1} is directed through R_{N1} and the parallel combination of C_N and R_{N2}. The voltage dropped across R_{N1} is a negative feedback voltage that is equal and opposite to that developed across the base-spreading resistance. In turn, the net voltage feedback to the input circuit is zero, and the amplifier has been unilateralized. Note that the values of C_N, R_{N1}, and R_{N2} depend upon the internal values of collector-base junction capacitance, base-spreading resistance, and collector resistance, respectively.

It is essential to make a careful worst-case analysis of the design for any unilateralized or neutralized RF amplifier. Otherwise, a large number of rejects may occur in production because of self-oscillation. In most cases, the designer makes an experimental analysis by assembling breadboard versions of the amplifier, using various combinations of selected components and devices that have both positive-tolerance limits and others that have negative-tolerance limits. As noted previously, it is not necessarily true that the worst-case situation will correspond to the occurrence of positive-tolerance limits throughout, or to the occurrence of negative-tolerance limits throughout. Often, the worst-case situation occurs when certain components and devices have their positive-limit tolerances, and the others have their negative-limit tolerances.

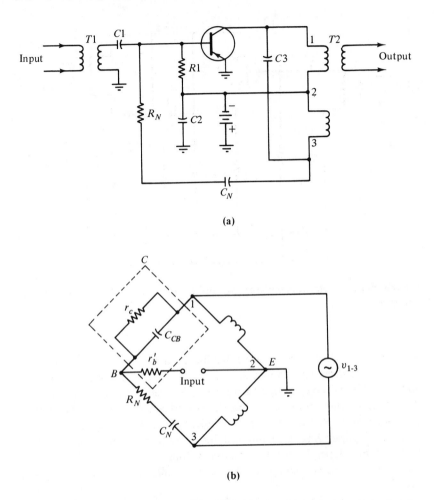

(a)

(b)

Figure 8-19 Common-emitter amplifier configuration, with bridge arrangement to prevent oscillation.

Common-emitter Amplifier; Bridge Unilateralization

A configuration for a common-emitter amplifier that is unilateralized by means of the transformer $T2$ winding 2–3 and the network consisting of R_N and C_N is depicted in Fig. 8-19. Transformer $T1$ couples the input signal to the base-emitter circuit. Transformer $T2$ winding 1-2 couples the output signal to the following stage. Resistor $R1$ forward-biases the transistor. Capacitor $C1$ prevents short-circuiting of the base-bias voltage by the secondary winding of $T1$. Capacitor $C2$ bypasses the collector battery and places terminal 2 of $T2$ at AC ground potential. Capacitor $C3$ tunes the primary of $T2$. Operation of this circuit is demonstrated to best advantage by means of the bridge arrangement formed by the transistor internal feedback elements and the external unilateralizing network as shown in Fig. 8-19(b).

The foregoing bridge arrangement does not involve $T1$, $C1$, $C2$, $C3$, $R1$, $T2$ secondary, or the collector battery. Points B, C, and E on the bridge represent the base, collector, and emitter terminals, respectively, of the transistor. Components 1, 2, and 3 correspond to the terminals of the $T2$ primary. The voltage that develops across terminals 1 and 3 of $T2$ is represented by a voltage generator with an output v_{1-3}. When the bridge is balanced, no part of voltage v_{1-3} appears between points B and E. In turn, the amplifier configuration is said to be unilateralized. The bridge is balanced when the ratio of voltages between points B and C and points B and 3 equals the ratio of the voltages between points C and E and points E and 3. In addition, the phase shift introduced by the network between points B and C must equal the phase shift introduced by the network between points B and 3.

8-8 LINE CHARACTERISTICS

A coaxial cable network for a CATV system must be carefully designed for maintenance of a fixed value of characteristic impedance (usually 75 ohms) throughout. Otherwise, system malfunctions owing to standing waves will occur. It is helpful to consider the nature of standing waves. With reference to Fig. 8-20, RF input voltage is applied at one end of a coaxial cable, and power is absorbed by a load resistor at the other end of the cable. A coaxial cable has a rated value of characteristic impedance. It is a basic electrical law that, if a transmission line is terminated in its own characteristic impedance, all of the power applied at the input end of the line will be absorbed by the load at the other end of the line. Moreover, the voltage is constant at any point along the line; that is, the voltage standing wave ratio (VSWR) is 1:1.

Next, consider the electrical relations in a transmission line that has no load at the far end, as depicted in Fig. 8-21. The power that flows into the cable proceeds at the speed of light to the far end, where it cannot be absorbed. Accordingly, this power is reflected from the open end of the cable back toward the input end. In its travel back, the reflected power mixes with the arriving power, and falls progressively out of phase until, at a point one-quarter wavelength back, the two power flows cancel each other. At this cancellation point, the line voltage is zero. This is called a *null point*. As the reflected power proceeds still farther back toward the input end, it then falls more nearly in phase

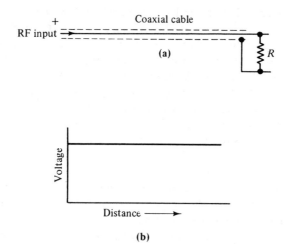

(a)

(b)

Figure 8-20 Transmission line terminated by resistor R. **(a)** Load absorbs the power flowing from the input end to the termination; **(b)** voltage along the line is uniform when R equals the characteristic impedance of the line.

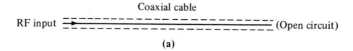

(a)

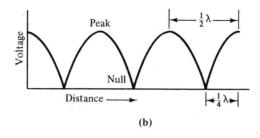

(b)

Figure 8-21 Example of complete mismatch, with far end of line open-circuited. **(a)** Power flowing from the input end cannot be absorbed at the far end of the line; **(b)** reflection of power from the far end produces nulls along the line.

with the arriving power. At a point one-half wavelength back, the two power flows become in phase with each other, and the line voltage is maximum at this point. This is called a *peak point*. The foregoing process is repeated at each half-wavelength interval back to the input end of the line. A line-voltage minimum is also called a voltage node, and a line-voltage peak is also called a voltage loop or voltage antinode. Observe that the VSWR is equal to the loop voltage divided by the node voltage; the value of the VSWR in the foregoing example is infinite.

Consider a situation in which the far end of a line is short-circuited, as depicted in Fig. 8-22. Since power cannot be absorbed by a short circuit, all of the power arriving at the far end of the line is reflected. This is similar to the reflection process described for an open-circuited line, except that a short-circuited termination produces a voltage zero at the short-circuit point. In turn, peaks and nulls occur at quarter-wavelength intervals, and the VSWR is infinite. Next, a condition of partial mismatch is shown in Fig. 8-23. In other words, the value of R is not equal to the characteristic impedance of the line. In this example, the value of R is less than the characteristic impedance value. This is shown by the fact that the voltage rises at first from the load toward the source. Observe that the maximum voltage on the line is equal to twice the minimum voltage. Therefore, the VSWR is 2:1.

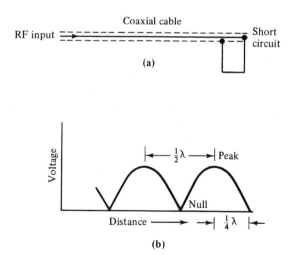

Figure 8-22 Example of complete mismatch, with far end of line short-circuited. (a) Power flowing from input end cannot be absorbed at the far end of the line; (b) reflection of power from the far end produces nulls along the line.

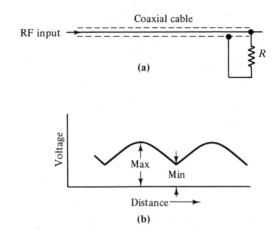

Figure 8-23   Example of partial mismatch. **(a)** Load resistor R is un-equal in value to the characteristic impedance of the line; **(b)** VSWR is 2:1.

It is instructive to follow the development of voltage and current standing waves at 45-deg intervals in time, as diagrammed in Fig. 8-24. Both voltage waves and current waves are present on the line. These waves fall progressively into phase and out of phase. When the current wave lags the voltage wave, the line has an inductive reactance at that point. When the current wave leads the voltage wave, the line has capacitive reactance at that point. An open-circuited line appears as an open circuit at each half-wave interval back from the load. At these points, it "looks like" a parallel-resonant circuit (see Fig. 8-25). A short-circuited line appears as a short circuit at each half-wave interval back from the load. At these intervals, it "looks like" a series-resonant circuit. These relations are summarized in Figure 8-26. Designers customarily utilize line sections instead of lumped inductance and capacitance for resonant circuits at ultra-high frequencies.

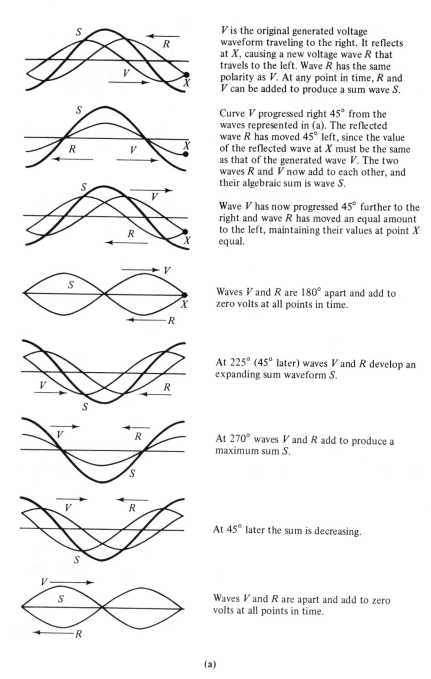

V is the original generated voltage waveform traveling to the right. It reflects at X, causing a new voltage wave R that travels to the left. Wave R has the same polarity as V. At any point in time, R and V can be added to produce a sum wave S.

Curve V progressed right 45° from the waves represented in (a). The reflected wave R has moved 45° left, since the value of the reflected wave at X must be the same as that of the generated wave V. The two waves R and V now add to each other, and their algebraic sum is wave S.

Wave V has now progressed 45° further to the right and wave R has moved an equal amount to the left, maintaining their values at point X equal.

Waves V and R are 180° apart and add to zero volts at all points in time.

At 225° (45° later) waves V and R develop an expanding sum waveform S.

At 270° waves V and R add to produce a maximum sum S.

At 45° later the sum is decreasing.

Waves V and R are apart and add to zero volts at all points in time.

(a)

Figure 8-24   How standing waves of voltage and current develop on an improperly terminated line. **(a)** Generated voltage waveform V, reflected voltage waveform R, and the sum of these waveforms S;

Since the current reverses at the open end of a line, the values of generated current *I* and reflected current *R* are opposite to those of the voltage waveforms *in* column (a).

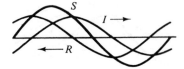

The generated wave *I* and the reflected wave *R* are in opposition, and their sum *S* is zero.

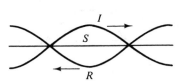

The generated wave *I* travels to the right, and the reflected wave *R* travels to the left. Their sum is represented by wave *S*.

The generated wave *I* and the reflected wave *R* produce a maximum sum *S*.

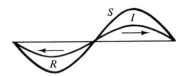

At 45° later the sum is decreasing.

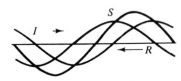

At 180° the generated wave *I* and the reflected wave *R* are out of phase, and their sum of zero.

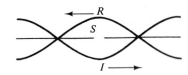

At 235° the sum *S* is increasing.

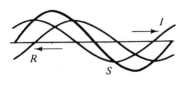

At 270° the generated wave *I* and the reflected wave *R* are again superimposed and develop a maximum sum *S*.

(b)

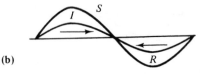

Figure 8-24 *Continued* **(b)** generated current waveform *I*, reflected current waveform *R*, and the sum of these waveforms *S*.

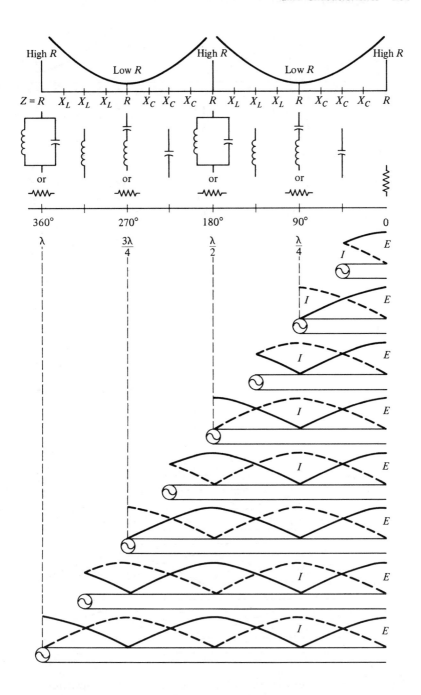

Figure 8-25   An open-ended line may "look like" an inductor, or a capacitor, or a resonant circuit.

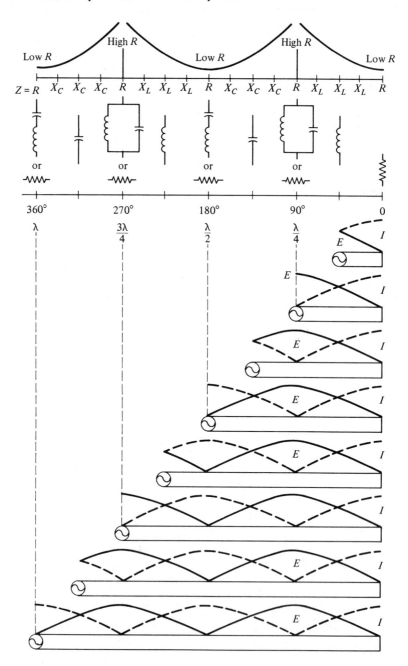

Figure 8-26   A short-circuited line may "look like" a capacitor, or an inductor, or a resonant circuit.

9

Production
Control Systems

9-1 GENERAL CONSIDERATIONS

Factory production control systems are characterized by automation technology. Automation is defined as a method of making a manufacturing or processing system partially or fully automatic. It encompasses the entire field of investigation, design, development, application, and methods of making processes or machines self-acting. Four basic factors are involved in factory production operations. These are the handling of material, processing of material, control of operations, and production design of the product. Electronics has provided new techniques of nondestructive testing, and new methods of control, with current emphasis on microprocessors and digital technology. Consider a piston factory in an automobile plant that is fully automated. Raw material in the form of metal bars is fed into an input device, and quality-controlled finished pistons are delivered by an output device

Automated processes are characterized in terms of processing, mechanical handling, sensing operations, and control operations. Processing involves the fabrication procedures that are required to convert raw material into a finished product. Mechanical handling denotes physical transport, locating, securing, and control of movements. Sensing is defined as automatic checking and correction (when required) of fabrication processes. As an illustration, a sensor might register a slight change in the diameter of a piston while it is being turned automatically in a lathe operation. Next, the sensor actuates a corrective mechanism that returns the cutting tool to its precise reference position. It follows that a corrective operation involves control means.

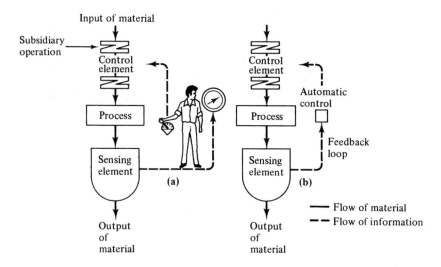

Figure 9-1 Control systems with human operator, and with automatic operation. **(a)** Control element manually operated; **(b)** control element automatically operated.

With reference to Fig. 9-1, manual operation of the control element is provided. In other words, a human operator observes an indicator and adjusts machine actions accordingly. On the other hand, automation employs a feedback loop for automatic operation of the control element. The two basic requirements involve the flow of material and the flow of information. Automatic control employs feedback from output to input to replace human control; the necessary information provided by the sensing element is converted into correct form for operation of the control element. The chief distinction between a mechanized system and an automated system is based on a feedback loop, as depicted in Fig. 9-2. Feedback loops often employ electronic components and devices. The program indicated in Fig. 9-2 may be as simple as a punched card or a punched tape (Fig. 9-3); again, the program may represent a large electronic computer.

A representative automated system includes a large number of action elements; these are of two general types: those designed for energy application, and those designed for transfer and positioning. Transfer and positioning elements typically control valves or conveyors, or otherwise transport and align the material being processed with respect to the elements for energy application that perform processing steps such as shaping, heating, or treating. The system may also include

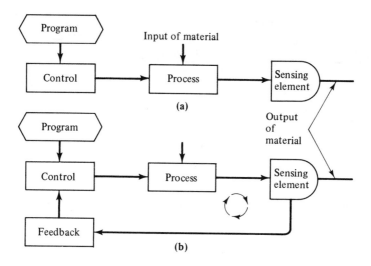

Figure 9-2 Comparative control systems. **(a)** Mechanized; **(b)** automated.

a large number of sensing elements. These elements detect and measure a specified property of the processed item; they convert the measurement data into a form that the automated system can act upon. Typical sensing elements include thermocouples, photocells, thermistors, gaussmeters, resistance bridges, strain gauges, and so on. An automated system may also include a large number of decision elements; these subsystems process information from sensing elements and compare it with that specified in the process program. Deviations result in the generation of instructions (typically electrical signals) for actuating the control elements. A decision element ranges in complexity up to a microprocessor or a large digital computer.

An automated system may also include a large number of control elements. These are typically mechanisms that carry out the instructions from the decision elements. The chief design problem is based upon the difficulty that is encountered (in many practical situations) in completely evaluating each of the action elements and element interactions, so that complete information may be entered into the process program.

Figure 9-3 Punched paper tape for process control.

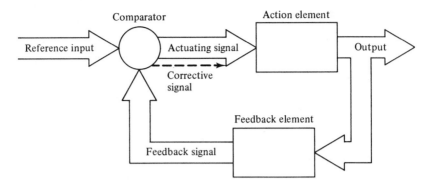

Figure 9-4   A simple action element with feedback facilities.

In other words, the designer may encounter puzzling debugging problems in his task of providing the proper decision-making routines at the required times without any hazard of false instructions or loss of control. A simple action element with feedback facilities is depicted in Fig. 9-4. The action element is completely responsive to the actuating signal; control is effected by comparison of the output from a feedback element with the reference input through a comparator. Any difference between the comparator inputs operates to modify the actuating signal until the difference is eliminated, or at least reduced to a tolerable amount.

9-2   STABILITY CRITERIA

The arrangement depicted in Fig. 9-4 is called a closed-loop system, because the output from the action element is eventually fed back to the input of the action element. Since it is a basic feedback system, it can develop instability with oscillation ("hunting"), or with overcorrection ("overshoot"). To avoid these malfunctions, the feedback phase must be negative under all conditions of operation. Any positive feedback can lead to transient or sustained abnormal output from the action element. The simplest way to consider stability criteria is to evaluate the performance of a basic amplifier with negative feedback. At midfrequency, the amplification is given by the equation

$$A_{fb} = A/(1 - \beta A)$$

where A is the amplifier gain without feedback.

A_{fb} is the amplifier gain with negative feedback.

β is the transfer coefficient; βA is the loop amplification.

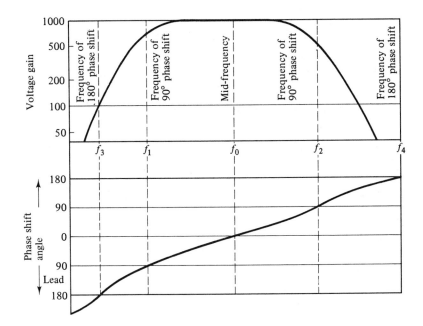

Figure 9-5   Frequency response and phase characteristic for a typical amplifier.

Both A and βA are vector (complex) quantities that have different amplitudes and different phase angles at very low frequencies and at very high frequencies, compared to their values at midfrequency. Negative feedback occurs when βA is real and negative; negative feedback results in reduced system amplification. On the other hand, if βA is real, positive, and less than unity in amplitude, positive feedback occurs with an increase in system amplification. Instability occurs if βA is equal to unity; this is a regenerative condition in which the amplification becomes infinite, at least in theory. Again, if βA is real, positive, and greater than unity in amplitude, the system becomes conditionally stable, and it may be thrown into oscillation by a suitable transient. This is the basis of the Nyquist stability criterion.

To evaluate an amplifier for stability, it is first necessary for the designer to determine its amplification and phase characteristics over the entire range of possible operating frequencies. For example, the frequency response in terms of voltage gain and phase shift versus frequency for a typical amplifier are depicted in Fig. 9-5. For a single-stage RC-coupled amplifier, a polar-coordinate plot of βA from zero to infinity gives the Nyquist diagram depicted in Fig. 9-6. The plot has cir-

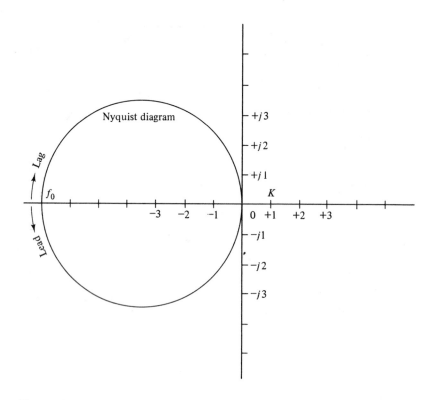

Figure 9-6 A Nyquist diagram for a single-stage RC-coupled amplifier.

cular form and is entirely in the negative region of the coordinates. Therefore, this amplifier is completely stable. If an amplifier has two stages, its Nyquist diagram does not plot as a circle. Parts of the diagram will extend into the positive region of the coordinates. Note point K. If an amplifier is stable, its Nyquist plot will not enclose point K; if the amplifier is conditionally stable, its Nyquist plot will not enclose point K, but the diagram will intersect the X axis at some point beyond K. Note that if the Nyquist diagram encloses the point K, the amplifier will be in a state of uncontrolled oscillation.

9-3 WARD-LEONARD SERVO DRIVE

It is frequently necessary to drive heavy rotary equipment at varying speeds and in either direction of rotation, and to maintain remote control of the drive. Since the ordinary single-phase AC motor is inherently a constant-speed device, as is the three-phase AC motor, designers generally utilize a DC motor for controlled drives. Its direction of rotation can be changed by reversing either the armature current, or the field current, but not both. The speed can be controlled by two chief methods; either the armature voltage can be varied, as depicted in Fig. 9-7(a), or the field voltage can be varied, as shown in Fig. 9-7(b). The first method provides the most stable control and the most desirable speed range. However, if a variable resistor is used in series with the armature, heavy power loss will occur, especially when larger sizes of motors are controlled. The second method permits control with a physically smaller resistor, and greatly reduces power losses, but the available motor speeds range from a certain minimum value upward, as indicated in Fig. 9-7(c). It is usually desirable to have the motor speed range from a certain maximum value to zero, as when the armature current is controlled. When the field is very weak, another disadvantage is that the motor becomes unstable in its higher speed ranges.

One design solution to this problem is to use the circuit depicted in Fig. 9-7(a), but to replace the resistor with a voltage-reducing device that has comparatively low power losses. The circuit shown in Fig. 9-8 is termed the Ward-Leonard system. This system provides the advantages of armature-current variation, while minimizing its disadvantages. The DC motor in this circuit is fed directly from a DC generator, which is operated at a constant speed. The DC field supply to the generator is variable in both magnitude and polarity by means of a rheostat and reversing switch, as shown. In turn, the motor armature is supplied by a generator that has smoothly varying voltage output from zero to full-load value. The motor field is supplied with a constant voltage from the same source that supplies the generator fields. The generator drive power could be from a single-phase or three-phase AC motor, or any other constant-speed source. In the same way, the DC field supply can be obtained from a rectifier, from an exciter on the end of the generator shaft, or other suitable DC source.

The chief advantage of the Ward-Leonard system is that by means of the variation of a small field current, a smooth and flexible yet stable control can be maintained over the speed and direction of rotation of a DC motor. The action of the system is very similar to that of an amplifier, in that a very small amount of power is used to control greatly increased power. A simple Ward-Leonard drive is depicted in Fig. 9-9. The DC generator in this system is driven by a 230-volt single-phase AC motor. The same AC line supplies a rectifier that furnishes field supply for both the DC generator and the DC motor fields. However, the generator field is connected to a potentiometer in such a way that the magnitude and polarity of the applied voltage can be varied. By

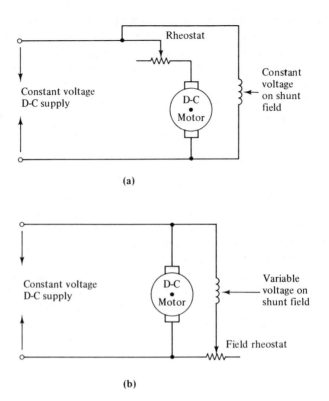

(a)

(b)

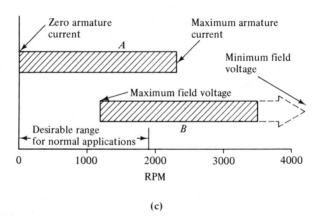

(c)

Figure 9-7   Basic speed control methods for DC motors. **(a)** Variable armature voltage-constant field flux; **(b)** constant armature voltage-variable field flux; **(c)** motor speed ranges.

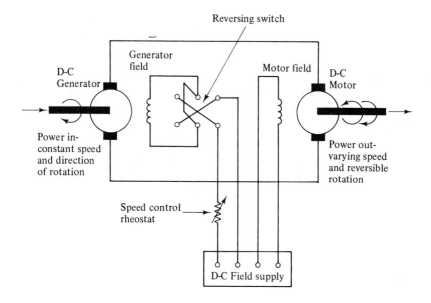

Figure 9-8 Basic Ward-Leonard drive.

varying the setting of the potentiometer control, rotation can be varied in either direction and at any speed from zero to full rate.

The basic Ward-Leonard system has certain limitations in some applications. For example, it may be desired to control the precise angular position of the rotatable shaft. It may be desired to rotate the shaft only a few degrees to another precise angular position. The speed of rotation may also need to be independent of varying resistance against the force of rotation. These requirements are met by employment of a servo system to supplement the basic drive system.

9-4 BASIC SERVO ARRANGEMENT

An adaptation of the Ward-Leonard drive to a servo system is shown in Fig. 9-10. The AC driving motor, DC generator, and the drive motor comprise the fundamental drive. The field supply for both the drive motor and the DC generator is obtained from an exciter, a small self-excited DC generator that is also driven by the AC driving motor. The field supply voltage for the DC generator is obtained from a Wheatstone bridge, comprising $R1$, $R2$, $R3$, and $R4$. If the resistance of $R1 = R3$, and $R2 = R4$, the bridge is balanced and the voltage between points A and B is zero. In this condition, the output from the DC generator

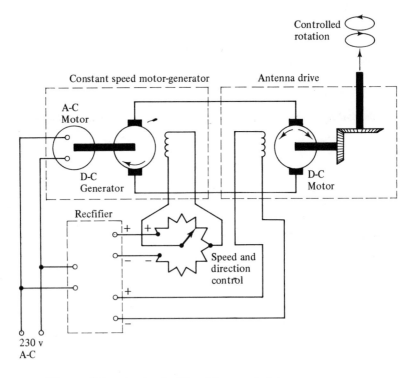

Figure 9-9   A simple Ward-Leonard drive arrangement.

is zero, so that the driving motor is stopped. However, if the resistance of $R1$ is decreased by short-circuiting part of the resistor, the bridge becomes unbalanced, making point A positive with respect to point B. The voltage across AB is impressed on the field of the DC generator, causing the drive motor to rotate. The speed at which the drive motor rotates depends on how much the resistance of $R1$ is reduced. Rotation in the opposite direction results from reducing the resistance of $R2$.

Resistors $R1$ and $R2$ are tapped at several points that are connected in turn to a pair of Silverstat contactors, as shown in the diagram. The resistance of either $R1$ or $R2$ is reduced by an amount dependent on how many contacts are closed. The Silverstats are operated by a lever arm L, which is geared to a differential selsyn motor. In turn, the speed and direction of rotation are determined by the lever arm that controls the voltage impressed on the DC generator field. When rotation is stopped, the rotor of DM1 is in a neutral position such that no contacts on either Silverstat are closed, and the Wheatstone bridge is balanced. If the handwheel on the rotor of the selsyn generator

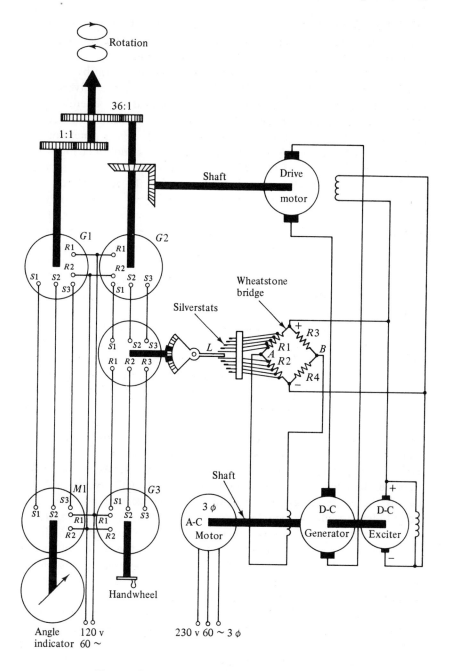

Figure 9-10 A Ward-Leonard servo system.

G3 is turned, the voltages induced in the stator will change. Since the rotor of the differential selsyn motor is energized from the stator of G3, the rotor of DM1 turns when the handwheel is turned.

Rotation of DM1 causes a deflection of the lever arm by the connection through bevel gears. This deflection causes the Silverstat to change the resistance of either R1 or R2 so that the DC generator produces a voltage that causes the drive motor to turn in the same direction that the handwheel was turned. Since the rotor of selsyn generator G3 is geared to the drive motor, the rotation of this motor changes the output from the stator of G2. This change in output causes a rotation of the stator field of DM1, which tends to bring the rotor of this machine to its neutral position. Thus, if the handwheel is turned only 10 deg, the rotating shaft will turn only 10 deg and stop. However, if the handwheel is turned continuously, the rotating shaft will turn continuously, because the rotor of DM1 is kept out of its neutral position by some constant small angle. This small angle, or error, is relatively large if the handwheel is turned rapidly, so that the rotating shaft also turns rapidly. On the other hand, the error is small if the handwheel is turned slowly.

The position of the rotating shaft may be indicated at a distance by the use of a selsyn system comprising G1 and M1. If desired, selsyn generator G1 may feed signals to several selsyn motors to indicate the rotating shaft position at several locations simultaneously. In a practical control system, hunting may occur because the inertia of the load may cause the control to overshoot the desired position. If this occurs, the voltages in the stator of G2 cause the lever arm to be displaced from its neutral position in a direction opposite to the displacement that caused the rotation. Overshooting causes the drive motor to reverse, and again the inertia of the system may cause it to overshoot, but in the opposite direction. Thus, overshooting develops a small error signal that causes the entire system to oscillate about the desired position. The amplitude of oscillation may be tolerable, but there is also a hazard in that the mechanical vibration that occurs may cause damage to the load.

For this reason, the lever arm that operates the Silverstat in the servo system may be modified by the addition of a gyroscope (Fig. 9-11) to reduce the hunting action. The gyroscope and its driving motor are mounted as a unit in trunnions on the lever arm. The position of the gyroscope relative to the lever arm is fixed in the static condition by a pair of balancing springs. Projection A on the gyroscope assembly operates the contacts of the Silverstat in a manner similar to the action of the lever itself. The rotor of the gyroscope is turned at high speed by the drive motor. Because of its great momentum, the gyro rotor tends to remain fixed in space. If the lever arm is turned about its pivot, the gyro tends to precess, or rotate, about the axis through the trunnions.

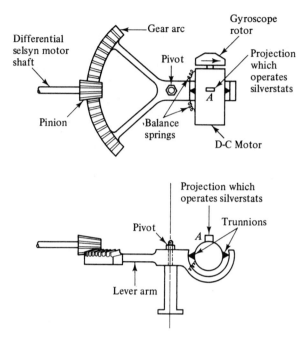

Figure 9-11   A gyroscopic antihunt arrangement.

In turn, the over-all action of the servo system is the same whether the gyroscope is operative or inoperative, but gyro action tends to eliminate hunting and to increase the speed of response to the controls. As an illustration, if the handwheel is turned suddenly, the rotor of DM1 immediately turns to a new position. The sudden motion of the lever arm moves A, which unbalances the Wheatstone bridge by operation of the Silverstat. In addition, the gyroscope precesses about its trunnions, causing a further motion of A in the same direction, which further unbalances the bridge. Thus, any sudden change of the controls causes high acceleration of the drive motor owing to the additive effect of gyro precession.

As the controlled shaft rotates, and the "error" in the field positions within DM1 decreases, the geared lever arm turns back toward its neutral position. The gyroscope is pulled to its normal position by the balancing springs, and it then precesses in the opposite direction as the lever arm moves toward neutral. This tilting reduces the unbalance of the Wheatstone bridge by pulling arm A back toward neutral more rapidly than would be possible by the lever arm alone, causing the drive motor to slow down as the shaft approaches its desired position.

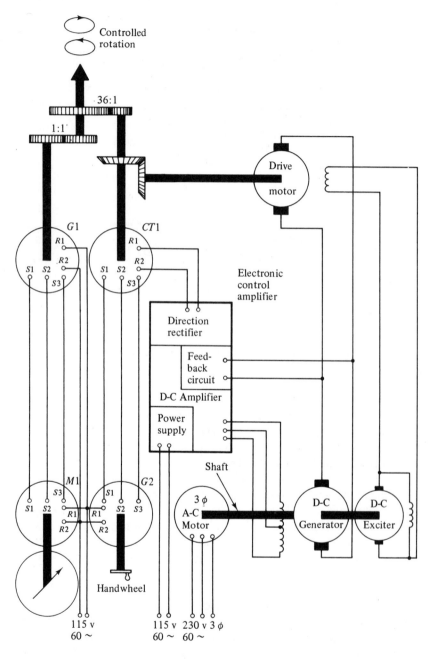

Figure 9-12   Electronic control of basic servo system.

This action tends to prevent the load on the shaft from overshooting or oscillating about the desired position. Thus, the action of the gyroscope is to cause the drive motor to be accelerated rapidly to start the load turning, and to cause the load to approach a desired position slowly to prevent hunting.

9-5 ELECTRONIC CONTROL OF BASIC SERVO SYSTEM

The foregoing control function may be performed electronically in a servo system. In this case, the differential selsyn, gyroscope, Silverstat, and Wheatstone bridge are replaced with a selsyn control transformer and an electronic amplifier that converts the selsyn AC voltage to direct current of sufficient amplitude to control the DC generator field in the drive. A skeleton diagram of this system is shown in Fig. 9-12. The position of the handwheel and rotor of the selsyn generator $G2$ determines the position of the field in the stator of the selsyn control transformer. No voltage is induced in the rotor of $CT1$ if it is at right angles to the stator-field flux. However, if the rotor is in any other position, a voltage is induced in its windings, which is fed to the control amplifier. The output from the amplifier causes the load to be rotated so that the rotor winding of $CT1$ is turned to a position at right angles to the stator field flux. In turn, voltage is no longer induced in the rotor of $CT1$, and the driving power is reduced to zero.

The direction rectifier is a circuit in which the phase of the error voltage output from the control transformer is compared with a reference voltage to determine the polarity of the control amplifier output. If the two voltages are in phase, the load is turned in one direction; if the two voltages are 180 deg out of phase, the load is turned in the other direction. The amplitude of the output from the control transformer determines the magnitude of the voltage applied to the field of the DC generator and, in turn, determines the speed of the drive motor. The feedback circuit functions to prevent hunting by reducing the amplifier gain as the voltage applied to the control motor decreases, thereby causing a further decrease of voltage applied to the motor. Thereby, the rate of load rotation is rapidly reduced as the desired position is reached, and the load is prevented from over-traveling.

9-6 OTHER SERVOMECHANISM ARRANGEMENTS

In addition to electric motor drive, factory production control systems employ other motive means, such as hydraulic pumps and controls, pneumatic circuits and boosters, gravity force, inertial force, and so on. These motive means are applied in handling, processing, and control operations. In their automated functions, each is integrated into a servo-mechanism—an automatic feedback control system in which one or more of its signals represent mechanical motion. In its broader aspect, each is an active component of a servo system in which the desired condition or value is compared with the prevailing condition or value, with development of an error signal for adjustment of the control element accordingly. In other words, any automated motive means comprises a reference input, comparator, action element with its output, and a feedback element. The same stability criteria apply, regardless of the motive means.

Integrated
Injection Logic (I²L)

One of the emerging integrated circuit technologies is integrated injection logic, or I²L. It combines two very desirable properties, simple fabrication and high-speed operation. The I²L devices are now the equal of saturated bipolar devices. I²L microprocessors are the beginning of a new family of IC circuits using this technology. The I²L gate is very simple. As shown in Fig. A-1-1, the basic gate structure is merely an inverter that is implemented as a multicollector transistor. A current source (a *pnp* transistor) supplies base drive to the *npn* multicollector injection element. The basic logic function is an inverter, and NAND functions can be provided with the basic inverter gate.

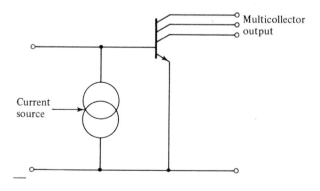

Figure A 1-1   Basic I²L gate structure.

One of the logic symbols used for I²L gates is shown in Fig. A-1-2. The individual multicollector outputs are shown as separate output lines, with a single input line to the inverter gate. Multiple inputs to the gate are wire-ANDed together (open collectors connected together) as depicted in Fig. A-1-2, and the inversion function of the multicollector transistor provides a NAND gate. The multicollectors provide separate output lines. The gate of Fig. A-1-2 has a "fan out" of three (capable of driving three additional gates). For a fan out of four or five, additional multicollector outputs would be added to the NOR inverter transistor.

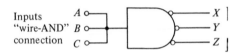

Figure A 1-2   A logic symbol used for I²L gates.

The basic I²L gate can be improved by the addition of Schottky clamp diodes on the output lines, as shown in Fig. A-1-3. The addition of the Schottky diodes on the output reduces the voltage swings and increases the operating speed. The simplicity of I²L gate fabrication is illustrated in Fig. A-1-4. The two transistors that form the gate are fabricated on a single N-type region. An additional P and N diffusion forms the multicollector transistor, and two P diffusions form the PNP current source transistor. The Schottky diodes are formed at the individual multicollectors. The oxide barrier shown in Fig. A-1-4 is used to separate the individual sections on the chip. The simplicity of the fabrication assures that the current distribution between the multicollector outputs is uniform. This eliminates the current-hogging condition

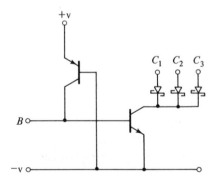

Figure A 1-3   Basic I²L gate with Schottky clamp diodes.

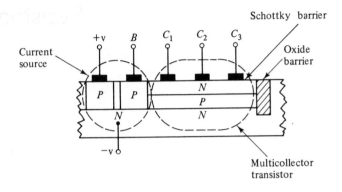

Figure A 1-4 Fabrication of an I²L gate.

that plagued early direct-coupled transistor logic systems. The I²L gate eliminates the need for resistors, thus saving chip space and increasing density.

The latch shown in Fig. A-1-5 illustrates the simplicity of I²L gate interconnections. Gates 1 and 2 are cross-coupled in the normal inverter fashion and together with the A and B inputs are wire-ANDed at the input. The basic gate structure eliminates connections within the logic element itself and greatly simplifies the inner connection of gate elements. A single circuit can operate over a wide speed range simply by varying the total current into the injection transistor. I²L densities range up to 250 gates per square millimeter. Schottky-clamped I²L gates can provide propagation delays in the one- to five-nanosecond range. Without Schottky diodes, propagation delays in the ten-nanosecond range can be obtained.

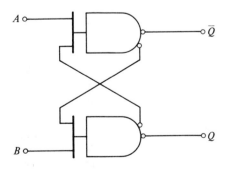

Figure A 1-5 Arrangement of an I²L latch.

Resistance

ELECTRONIC INDUSTRIES ASSOCIATION (EIA) PREFERRED VALUES

Commercial tolerances on resistance values are 5, 10, and 20 percent. EIA preferred values are nominal resistance values that avoid duplication of stock within each tolerance range. The nominal resistance values tabulated below can be multiplied or divided by any power of 10.

TOLERANCE

5%	10%	20%	5%	10%	20%
10	10	10	36		
11			39	39	39
12	12				
13			43		
15	15	15	47	47	47
16			51		
18	18		56	56	
20			62		
22	22	22	68	68	68
24			75		
27	27		82	82	
30			91		
33	33	33	100	100	100

Color Codes

RESISTOR COLOR CODES

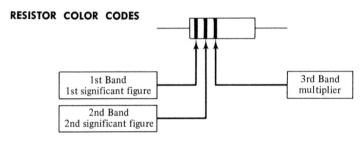

3-band code for ±20 percent resistors only

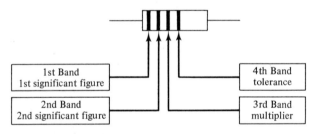

4-band code for resistors with ±1 percent to
±10 percent tolerance

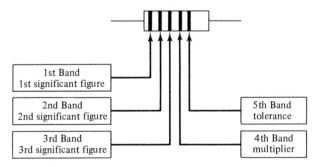

5-band code for resistors with ±1 percent to
±10 percent tolerance

Figure A III-1

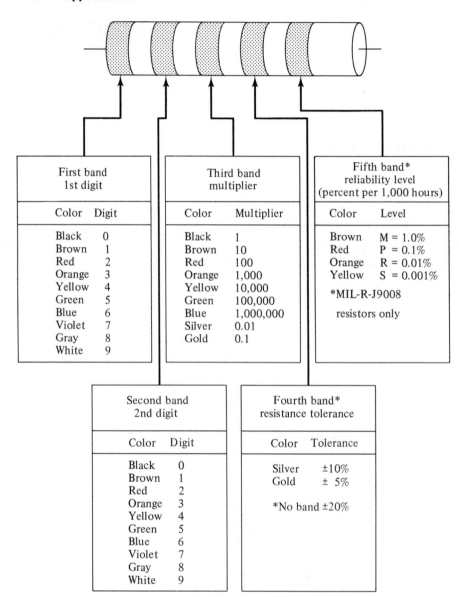

First band 1st digit		Third band multiplier		Fifth band* reliability level (percent per 1,000 hours)	
Color	Digit	Color	Multiplier	Color	Level
Black	0	Black	1	Brown	M = 1.0%
Brown	1	Brown	10	Red	P = 0.1%
Red	2	Red	100	Orange	R = 0.01%
Orange	3	Orange	1,000	Yellow	S = 0.001%
Yellow	4	Yellow	10,000		
Green	5	Green	100,000	*MIL-R-J9008	
Blue	6	Blue	1,000,000		
Violet	7	Silver	0.01	resistors only	
Gray	8	Gold	0.1		
White	9				

Second band 2nd digit		Fourth band* resistance tolerance	
Color	Digit	Color	Tolerance
Black	0	Silver	±10%
Brown	1	Gold	± 5%
Red	2		
Orange	3	*No band ±20%	
Yellow	4		
Green	5		
Blue	6		
Violet	7		
Gray	8		
White	9		

5-band code for resistors with ±1 percent to ±10 percent
tolerance (fifth band reliability level)

Figure A III-2

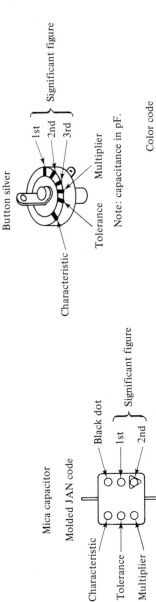

Mica capacitor color code

Mica capacitor

Molded JAN code

Retma code

Note: capacitance in pF. If both rows of dots are not on one face, rotate capacitor about the axis of its leads and read second row on side or rear.

Color code

Color	Significant figure	Multiplier	Tolerance %	Characteristic
Black	0	1	20	A
Brown	1	10		B
Red	2	10^2	2	C
Orange	3	10^3	3(RETMA)	D
Yellow	4	10^4		E
Green	5		5(RETMA)	F (JAN)
Blue	6			G (JAN)
Violet	7			—
Gray	8			I (RETMA)
White	9			J (RETMA)
Gold	—	0.1	5(JAN)	—
Silver	—	0.01	10	—
None	—	—	20(RETMA)	—

Figure A III-3

275

Paper capacitor color code

JAN code

Characteristic
Tolerance
Multiplier

Silver dot
1st } Significant figure
2nd } figure

Color	Significant figure	Multiplier	Tolerance %	Characteristic
Black	0	1	20	A
Brown	1	10	–	B
Red	2	10^2	–	C
Orange	3	10^3	30	D
Yellow	4	10^4	40	E
Green	5	10^5	5	F
Blue	6	10^6	–	G
Violet	7	–	–	–
Gray	8	–	–	–
White	9	–	10	–
Gold	–	1	5	–

Note: capacitance in picofarads. Working voltage coded in terms of hundreds of volts. Voltage ratings over 900 V expressed in two-dot voltage code.

Molded flat RETMA code

Multiplier
2nd } Significant figure
1st } figure

Working voltage
Black body

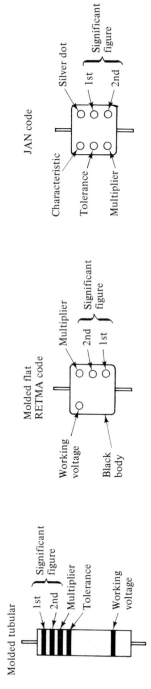

Molded tubular

1st } Significant figure
2nd } figure
Multiplier
Tolerance
Working voltage

Color	Significant figure	Multiplier	Tolerance %
Black	0	1	20
Brown	1	10	–
Red	2	10^2	–
Orange	3	10^3	30
Yellow	4	10^4	40
Green	5	10^5	5
Blue	6	10^6	–
Violet	7	–	–
Gray	8	–	–
White	9	–	10
Gold	–	1	–

Note: capacitance in picofarads. Working voltage coded in terms of hundreds of volts. Ratings over 900 V expressed in two-band voltage code.

Figure A III-3 Continued

Ceramic capacitor

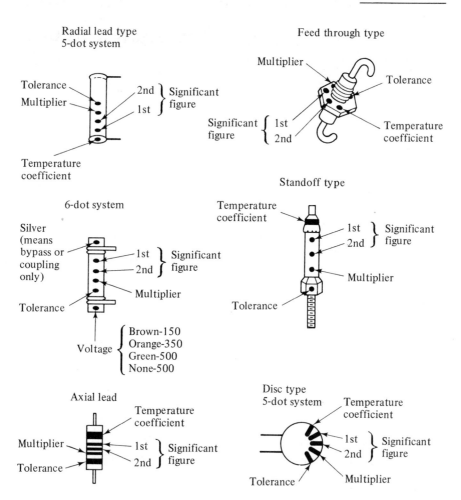

Radial lead type
5-dot system

Tolerance
Multiplier
2nd ⎱ Significant
1st ⎰ figure
Temperature
coefficient

Feed through type

Multiplier
Tolerance
Significant ⎰ 1st
figure ⎱ 2nd
Temperature
coefficient

6-dot system

Silver
(means
bypass or
coupling
only)
Tolerance
1st ⎱ Significant
2nd ⎰ figure
Multiplier
Voltage ⎰ Brown-150
⎱ Orange-350
Green-500
None-500

Standoff type

Temperature
coefficient
1st ⎱ Significant
2nd ⎰ figure
Multiplier
Tolerance

Axial lead

Temperature
coefficient
Multiplier
1st ⎱ Significant
2nd ⎰ figure
Tolerance

Disc type
5-dot system

Temperature
coefficient
1st ⎱ Significant
2nd ⎰ figure
Tolerance
Multiplier

Figure A III-4

Characteristics of Series- and Parallel-Resonant Circuits

Quantity	Series Circuit	Parallel Circuit
At resonance: Reactance $(X_L - X_C)$	Zero; because $X_L = X_C$	Zero; because nonenergy currents are equal
Resonant frequency	$\dfrac{1}{2\pi\sqrt{LC}}$	$\dfrac{1}{2\pi\sqrt{LC}}$
Impedance	Minimum; $Z = R$	Maximum; $Z = \dfrac{L}{CR}$, approx.
I_{line}	Maximum	Minimum value
I_L	I_{line}	$Q \times I_{line}$
I_C	I_{line}	$Q \times I_{line}$
E_L	$Q \times E_{line}$	E_{line}
E_C	$Q \times E_{line}$	E_{line}
Phase angle between E_{line} and I_{line}	$0°$	$0°$
Angle between E_L and E_C	$180°$	$0°$
Angle between I_L and I_C	$0°$	$180°$
Desired value of Q	10 or more	10 or more
Desired value of R	Low	Low
Highest selectivity	High Q, low R, high $\dfrac{L}{C}$	High Q, low R
When f is greater than f_0: Reactance	Inductive	Capacitive
Phase angle between I_{line} and E_{line}	Lagging current	Leading current
When f is less than f_0: Reactance	Capacitive	Inductive
Phase angle between I_{line} and E_{line}	Leading current	Lagging current

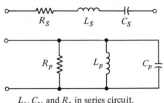

L_s, C_s, and R_s in series circuit.
L_p, C_p, and R_p in parallel circuit.

$$Q = \frac{\omega L_s}{R_s} = \frac{1}{\omega C_s R_s} = \frac{R_p}{\omega L_p} = R_p \omega C_p = \frac{\sqrt{L_s/C_s}}{R_s} = \frac{R_p}{\sqrt{L_p/C_p}}$$

General Formulas	Formulas for Q Greater than 10	Formulas for Q Less than 0.1
$R_s = \dfrac{R_p}{1 + Q^2}$	$R_s \simeq \dfrac{R_p}{Q_2}$	$R_s \simeq R_p$
$X_s = X_p \dfrac{Q^2}{1 + Q^2}$	$X_s \simeq X_p$	$X_s \simeq X_p Q^2$
$L_s = L_p \dfrac{Q^2}{1 + Q^2}$	$L_s \simeq L_p$	$L_s \simeq L_p Q^2$
$C_s = C_p \dfrac{1 + Q^2}{Q^2}$	$C_s \simeq C_p$	$C_s \simeq \dfrac{C_p}{Q^2}$
$R_p = R_s (1 + Q^2)$	$R_p \simeq R_s Q^2$	$R_p \simeq R_s$
$X_p = X_s \dfrac{1 + Q^2}{Q^2}$	$X_p \simeq X_s$	$X_p \simeq \dfrac{X_s}{Q^2}$
$L_p = L_s \dfrac{1 + Q^2}{Q^2}$	$L_p \simeq L_s$	$L_p \simeq \dfrac{L_s}{Q^2}$
$C_p = C_s \dfrac{Q^2}{1 + Q^2}$	$C_p \simeq C_s$	$C_p \simeq C_s Q^2$
$B_L = \dfrac{1}{X_L}$	$B_L = \dfrac{1}{X_L}$	$B_L = \dfrac{1}{X_L}$
$B_c = \dfrac{1}{X_c}$	$B_L = \dfrac{1}{X_L}$	$B_L = \dfrac{1}{X_L}$
$Y = \sqrt{G^2 + B^2}$	$Y = \sqrt{G^2 + B^2}$	$Y = \sqrt{G^2 + B^2}$

Index

AB operation, 206
A&D converter, 124
Accumulator, 141
Active
  signal, 159
  tone control, 200
Activity, system, 165
Address, 141
  access, 143
  flow, 161
  keyboard, 153
  latches, 150
Adjacent channel, 57
Adjustable bandwidth, 103
Aliasing, 116
Allocated channel, 49
Amplification, 100
Amplifier, audio, 181
Amplitude
  modulation, 127
  nonlinearity, 78
Analyzer, digital, 150
AND gate, 134
Antenna, radar, 100
Aperture radiators, 64
Arm, 144
Assembler, cross, 153

Assignments, pin, 147
Audio
  amplifier, 181
  distortion, 198
  power, 179
  signals, 63
Automatic
  feedback control, 268
  volume control, 52
Automation, 253
Autotransformer, 108
AVC, 52

Balanced modulator, 38, 46, 59
Balancing networks, 35, 69
Bandpass filter, 40, 48, 115, 217
Beam, directional, 92
Beta value, 196
Bias, 242
Binary
  code, 112
  digit, 28
  pulse code, 115
Binaural perception, 58
Bit, 143

Black
  compression, 87
  stretch, 87
  transfer characteristics, 87
Blanking, 147
  signal, 78
  vertical, 25
Bogie value, 185
Booth, control, 61
Bridge transformer, 32, 68
Bridged
  equalizer, 21
  T arrangement, 21
Broadcast
  band, 49
  network, 49, 62
  system, 50
Byte, 147

Cable
  carrier systems, 66
  connectors, 26
Carrier
  repeater, 66
  systems, 42, 64
  telephony, 38
  transmission, 86
CATV, 217
Central processing unit, 135
Channel allocations, 51
Characteristic
  impedance, 13, 37
  resistance, 13
Chroma phase, 26
Chrominance signals, 78
Clear channel, 55
Clock, digital, 135
Coaxial cable, 15
Co-channel interference, 55
Code pulse, 125, 133
Coincidence switch, 134
Color burst, 25
Communication circuit, 4
Community antenna, 217
Commutation, 124

Commutation (*Cont.*)
  duty cycle, 124
  frame period, 24
Composite
  audio signal, 59
  color signal, 78
  telemetry signal, 129
Control
  booth, 61
  room, 63
CPU, 135
Critical angle, 89
Crossover
  circuit, 210
  distortion, 206
Crosstalk, 23, 33, 36, 134
Crystal mixer, 106
Cutoff
  frequency, 34
  point, 84

Darlington connection, 206
Data storage, 136
dBm, 63
Debugging, 138, 256
Decoder, 113
Decommutator, 124
Delay
  adjustment, 134
  line, 13
  multivibrator, 134
  time, 70, 81
Demodulator, 38
Depletion FET, 192
Design compatibility, 164
Detector, envelope, 60
Differential
  gain, 86
  phase, 86
Digital
  logic, 134
  pulses, 30
Digitizer, 124
Dipole, 100
Direct wave, 51

Directional
  beam, 92
  transmission, 64
Discriminator
  pulse-width, 132
  subcarrier, 123
Display indicator, 91
Distortion, harmonic, 198
Distributed
  capacitance, 15
  inductance, 15
Diversity
  gain, 27
  reception, 27
Double sideband, 81

Echo
  chamber, 63
  pulse, 103
  signal, 89
Effective stabilization, 198
Efficiency, 204, 212
  operating, 218
  system, 219
EHF, 92, 121
Eight-track deck, 178
Electromagnetic waves, 100
Electronic control, 267
Enclosure, speaker, 179
Encoder, 124
Energy
  application, 254
  signal, 217
Engineering, system, 1
Envelope detection, 60
Equalization, 72
Equalizer, 18, 239
Equivalent circuit, 106, 201, 227
Error, tolerance, 224
Exciter waveform, 127
Execution, subroutine, 171
Extended response, 205
Extender, line, 217
External feedback, 240

Fader, 63
Fading, 52
Fall time, 21
FCC, 49
Feedback, 192
  current, 197
  loop, 200
  negative, 196
  significant, 196
  voltage, 197
Feeder line, 218
Feedthrough, 44
FET, 191
  depletion, 192
  enhancement, 192
Fetch, digital, 153
  cycle, 173
Fidelity, 202
Field-effect transistor, 191
Film camera, 74
Filter, 181
  bandpass, 217
  RC, 202
Filters, 37
Flat-topped response, 236
FM
  decoder, 115
  system, 57, 111
  tuner, 178, 183
Forward bias, 242
Four-wire repeater, 69
Frequency
  beta cutoff, 196
  diversity system, 27
  division multiplexing, 40
  equalizer, 20
  function, 187
  modulation, 57
  range, 181
  response, 182, 196, 202, 214
  selective circuitry, 200
  sharing system, 115
  shift keying, 38
Fringe areas, 217
FSK, 124

Gauge, strain, 122
Gain
    audio amplifier, 181
    antenna, 217
    current, 184
    differential, 86
    diversity, 26
    maximum
        available (MAG), 182
        usable (MUG), 36
    minimum, 196
    power, 184
    voltage, 184
Gamma, 25, 72
Gate
    AND, 134
    multivibrator, 134
    NAND, 140
    pulses, 128, 134
    starter, 130
Gaussmeter, 255
Generator
    resistance, 188
    voltage, 195
Germanium transistor, 187
Ground
    currents, 24
    reflected wave, 52
    wave, 51
Guard band, 40
Gyroscope, 264

h.a.d., pulse, 78
Half
    amplitude duration, 80
    power
        frequencies, 17
        points, 110
Harmonic
    distortion, 181, 198
        percentage, 200
Harmonics, 48
Head end, 217
High

High (*Cont.*)
    fidelity, 178
        amplifiers, 181
        range, 178
        speakers, 214
        systems, 178
    frequency response, 200
    gain antenna, 217
    impedance, 184
    pass filter, 78
Horn-type tweeter, 213
Hue, 88
Hum
    nonsynchronous, 76
    synchronous, 76
    voltage, 23
Hunting, 256
Hybrid coil, 32, 37, 68
    subsystem, 69

IC, 135
IF amplification, 103
IGFET, 192
I²L logic, 269
Impedance
    characteristic, 13, 37
    relations, 98
    surge, 13
    transfer, 9
Index
    modulation, 120
    register, 136
Indicator
    display, 91
    radar, 89
Indirect wave, 52
Inductance, distributed, 15
Induction coil, 32
Information
    bits, 28
    processor, 113
    transfer, 28, 116, 131
Integrated
    circuit, 135
    injection logic, 269

Intelligibility, 56
Interchannel crosstalk, 120
Interference, 49
   co-channel, 55
Intermittent service, 53
Interstage coupling, 108
I/O, 171
Ionosphere, 54, 89
Irregularities, phase, 81

JCN, 141
JFET, 192
Jump, conditional, 143
JUN, 162
Junction
   capacitance, 218, 228
   T, 99
   transistors, 184

Keyboard, 153
Klystron oscillator, 127
kQ product, 237

Lagging edge, 84
Latch, digital, 271
Leading
   edge, 84
   lobe, 83
Line
   amplifier, 70, 74
   current, 221
   extender, 217
Linear distortion, 72
Load impedance, 183
Loading, 34
   coils, 34
Lobes, 83
Local channel, 55
Logic
   circuit, 28
   integrated injection, 269
   negative, 168
Lower sideband, 144

Low-pass filter, 40
LSB, 144
Luminance signal, 78

MAG, 182
Marker
   generator, 126
   pulse, 125
Maximum
   available gain, 182
   usable gain, 36
Memory, digital, 139
Microprocessor, 135
   organization, 137
Microwave, 61, 100
Mismatch, 100
Mixer cavity, 108
Modem, 42
Modulation
   index, 120
   spectrum, 58
Modulator
   balanced-bridge, 42
   suppressed carrier, 42
Monitor oscilloscope, 78
Monochrome transmission, 78
Monophonic signal, 61
MSB, 150
MUG, 37
Multiburst signal, 7
Multichannel operation, 40
Multiplex telemetry, 132
Multiplexing, 30, 58, 112, 118, 137
   frequency-sharing, 30
   time-division, 30
Multivibrator, 130

NAND gate, 140
Negative
   feedback, 74
   logic, 168
Network
   analysis, 7
   analyzer, 13

Network (*Cont.*)
  balancing, 35, 69
  broadcast, 49, 62
  channels, 54
  crossover, 210
  lines, 64
  television, 23
  transcontinental, 86
  video, 70
Noise, 4, 196
  factor, 4
  level, 5, 102
  partition, 102
  system, 1
  voltages, 102, 133
Nonlinear
  amplifier, 87
  distortion, 72, 85
  resistance, 44
Nonlinearity
  amplitude, 78
  residual, 186
Nonsynchronous hum, 76
Null point, 245

One-way repeater, 66
Open
  circuited line, 247
  wire line, 40
Operand, 153
Operation
  multichannel, 40
  phantom, 33
Operational checks, 84
Optimum Q, 236
Organization, MP, 137
Oscillation, 35, 240
Oscillator
  klystron, 127
  self-pulsing, 95
  timing, 96
Oscilloscope, 78
Output
  impedance, 228
  signal, 236

Overshoot, 81, 256, 264
Oxide barrier, 270

PAM, 111
Parallel resonance, 278, 279
Parity, 28
Partition noise, 102
Patch panel, 123
PCM, 111
PDM, 111
Phantom
  circuits, 32
  operation, 33
Phase
  angle, 17
  characteristic, 17, 21, 72
  distortion, 21
  equalization, 72
  equalizer, 20
  irregularities, 81
  relations, 12, 53
  response, 83
Polarization, 64
Port, 141
PPM, 111
Preamplifier, 63
Predistortion, 25, 72, 81
Preferred values, 272
Primary service area, 51
Probe, 108
Program listing, 141
Protective device, 97
Proximity device, 123
Pulse
  amplitude, 85, 127
  converter, 131
  multiplexing, 118
  time, 97
  width, 131

Quadded conductors, 33
Quadraphonic, 61, 178
Qualifiers, 144
Quality factor, 221
Quantizing section, 115
Q value, 221

Radar, 89
  antenna, 100
  indicator, 91
  system, 92
  timer, 92
Radio
  broadcast system, 50
  relay net, 72
RAM, 135
Ramp, 78, 130
RC filter, 202
Real time, 138
  monitoring, 123
Receiver
  bandwidth, 103
  subsystem, 109
Rectangular pulses, 103
Redundancy, 26, 135
Reflection coefficient, 37
Regenerated pulse, 130
Regional channel, 55
Reliability, 26, 135
Repeater stations, 64
Repeaters, 34, 37
Repeating coils, 33
Repetition rate, 97
Reset, 136
Residual
  frequency drift, 103
  nonlinearity, 86
Resonant cavity, 98, 103
Resultant wave, 52
Return loss, 37
Ringing, 83
Rise time, 21, 119, 127
ROM, 136

Sampling
  process, 115
  rate, 116
SCA, 59
Self-oscillation, 35
Sensors, 121
Series resonance, 278, 279
Service area, 49

SHF, 64
Sideband, 22
Sidetone, 32
Signal
  conditioners, 121
  to noise ratio, 5, 100, 103
Single sideband, 38
Sine-squared pulse, 80
Space diversity, 27
Spiral four quad, 33
Square wave, 18
SSB, 38
Stabilizing amplifier, 74
Staircase waveform, 78
Standardized level, 5
Standing-wave ratio, 26
Static, 23
Step response, 83
Stereo decoder, 59
Stereophonic sound, 58
STL, 61
Strain gauge, 122
Subcarrier, 59
  source, 113
Subsystems, 4
Surge impedance, 13
Synchronous hum, 76
Sync pulse, 8
System
  deviation, 1
  engineering, 1
  noise, 1
  reliability, 1, 26

Telco, 61
Telemeter, 111
Telemetry, 111
  band, 121
  frame, 112
  links, 112
  system, 116
Telephone
  cables, 64
  circuits, 30
Television network, 23

Temperature compensation, 74
Ternary PCM, 116
Thermal agitation, 102
Tilt, 21, 81
Time delay, 17
Time
    constant, 134
    diversity reception, 27
    multiplexed signals, 61
Timing pulse, 94
T junction, 99
Tolerances, 23
T pulse, 84
Trailing lobe, 83
Transcontinental networks, 86
Transducer, 112
Transfer
    admittance, 9
    characteristic, 87
    impedance, 9
Transient response, 72
Transit time, 70
Transmission line, 15, 103
TR switch, 98
TTL, 148
Tuned filter, 66
TV camera, 74
Two-way repeater, 66

UHF, 217
Uncontrolled oscillation, 240
Undershoot, 81
Unilateralization, 240
Unique
    clock, 161
    trigger, 144
Unitized hi-fi, 178
Unity
    coupling, 229
    gain, 200

Valve, control, 254
Varistor, 40
Vectorscope, 87

Vertical blanking, 25
Vestigial sideband, 25, 72
VHF, 217
    omnirange, 57
Video network, 70
VIR, 25
VITS, 77, 81
Voice currents, 30
Volume unit, 63
VOR, 57
Voltage
    amplification, 189
    feedback, 196
    gain, 183
    operated device, 192
Volume control, 203
VSB, 25, 72
VSWR, 26, 245
VU, 63

Waveform
    complex, 200
    distortion, 200
Wavelength, 92
Wheatstone bridge, 44, 265
White
    compression, 87
    stretch, 87
Wide-band response, 228
Window pulse, 81
Woofer, 178, 210
Worst-case
    condition, 7
    principles, 204

X-Y plotter, 112

Zero
    bias configuration, 205
    emitter-base voltage, 204
    state display, 155